Hydrogen Molecular Biology and Medicine

Xuejun Sun • Shigeo Ohta • Atsunori Nakao
Editors

Hydrogen Molecular Biology and Medicine

Springer

Editors
Xuejun Sun
Department of Navy Aeromedicine
Second Military Medical University
Shanghai
China

Atsunori Nakao
Department of Emergency
Hyogo College of Medicine
Nishinomiya
Japan

Shigeo Ohta
Department of Biochemistry and Cell Biol
Nippon Medical School
Kawasaki
Japan

ISBN 978-94-017-9690-3 ISBN 978-94-017-9691-0 (eBook)
DOI 10.1007/978-94-017-9691-0

Library of Congress Control Number: 2015933014

Springer Dordrecht Heidelberg New York London

Printed on acid-free paper

Springer is part of Springer Science+Business Media (www.springer.com)

Preface

Oxygen sustains the lives of thousands of aerobionts as an irreplaceable element. Oxygen is necessary and beneficial. It is the particularly final electron acceptor in human body. In cells, the components of the entire human entity, oxygen assists the continual process of metabolism of substances and energy. In other words, energy that sustains all the life activity of the body comes from nutritive substances that cannot be in occupation without the existence of oxygen. However, oxygen is unnecessary and harmful when it is excessive or over tension. People who breathe oxygen that is over 50 % of the volume fraction of air for a considerable period may have problems. The excess of oxygen may convey toxicity, even causing pulmonary injury which is also defined as pulmonary oxygen intoxication, demonstrating intoxication caused by oxygen overdose inhalation. In addition, breathing hyperbaric oxygen that exceeds 304 kPa (3 atm) could introduce encephalon oxygen poisoning. For now, it is believed that oxygen toxicity is caused by excess oxygen introducing reactive oxygen species (ROS) in the organism.

ROS has a long history of accusations of culpability in different kinds of diseases, and it is convincible that eliminating ROS will cure the disease. However, this viewpoint is not partial but unacceptable. Although ROS over a certain degree causes body damage, ROS of normal concentration is an irreplaceable condition for maintaining health. Why is ROS important for the health of the body?

ROS is derived of free radical oxygen or non-free radical oxygen with intensely oxidative or reductive functions. We can also put it this way, that ROS is an oxygen element-containing group with powerful reactive activity in aerobic organisms, such as human beings. The most common ROS in the body is superoxide anion, hydrogen peroxide, and nitric oxide. Nitric oxide is one of the crucial signal molecules with an extensive biological effect. Vascular endothelial cells rely on nitric oxide to modulate normal blood pressure. In general, nitric oxide is one kind of significant free radical and is also a major kind of ROS.

Therefore, it is a common misunderstanding that ROS are destructive to the human body and it is better to remove them to protect the body. Considering all ROS as a package and dealing with all of them is the wrong idea. There are a variety of ROS in organism. One group, such as superoxide anion, hydrogen peroxide, nitric oxide etc., has relatively low activity and is favorable to the human body. Only a situation where these ROS are produced excessively could cause damage to body. Another group, such as hydroxyl free radical, nitrous acid anions, and hypochlorous acid radical, is highly reactive, although these free radicals are rare, reacting easily with intracellular protein, nucleic acid, and lipids by irreversible chemical reaction. These reactions cause fatal damage to the molecule. This damage is called oxidative damage. This kind of injury is the most common and basic pathophysiology mechanism for generation and development of human disease. There is a certain amount of clinical and basically medical research that has verified that common diseases, such as cardiovascular disease, cerebrovascular disease, inflammation, malignant carcinoma, diabetes and arteriosclerosis are caused by ROS introduced oxidative damage. Essentially, it is the ROS group with a minor amount but intoxicate capacity that develops oxidative damage. In general, there are two kinds of ROS. One is the ROS group characterized as large in number, low in activity and beneficial; the other ROS group is small in number, high in activity, and destructive to the body. According to this, there are flaws in treating ROS-caused disease by administrating strong reductive drugs (as vitamins) to eliminate all ROS. A strong reductant treats ROS as a whole and cannot selectively remove those that are high in reactivity and harmful. Only substance with a selectively antioxidant effect (which selectively neutralizes harmful ROS) is the perfect antioxidative drug. It has been demonstrated by a lot of research, that hydrogen could selectively neutralize hydroxyl free radicals and nitrous acid anions. This is the fundamental theory for hydrogen treating disease via its antioxidative effect.

According to the research, hydrogen shows a protective effect in multiple diseases. For instance, malignant carcinoma, colitis, encephalopathia after carbon monoxide poisoning, cerebral ischemia, senile dementia, Parkinson's disease, depression, spinal injury, skin allergy, diabetes type 2, acute pancreatitis, organ transplantation, intestinal ischemia, systematic inflammation reaction, radioactive injury, retina injury, deafness, etc. However, till now, only type 2 diabetes, cerebral ischemia, rheumatoid arthiritis, and Parkinson's disease are in the process of clinical trials. Others still need rigorous human tests to confirm the effects.

While hydrogen has potential value for tackling disease, the method of application is a challenge we have to face to employ its antioxidant effect properly. There are three kinds of hydrogen application. One method is breathing a mixture of hydrogen and oxygen directly; another method is taking in hydrogen through digestion or the venous system; the third method is spreading hydrogen subcutaneously or introducing aerogenic bacteria to generate hydrogen. It is much more practical to employ the method of drinking hydrogen water for people from the point of view of both efficiency and affordability. Hydrogen related products have been invented by many companies and sold generally to areas such as Japan and China. We truly believe that hydrogen research and hydrogen related products will enlighten us to a brighter, more intellectual, and healthier way of living.

May all the goodness live longer!

Department of Navy Aeromedicine, Dr. Xuejun Sun
Second Military Medical University

Contents

Contributors

Dong Cao Department of Osteology, The sixth people's hospital, Shanghai, China

Xiao Chen Department of Orthopedics trauma, Shanghai Changhai Hospital, Second Military Medical University, Shanghai, China

Wenjie Han Department of Cadre Ward, Navy General Hospital, Beijng, China

Wenwu Liu Department of Diving Medicine, Second Military Medical University, Shanghai, China

Atsunori Nakao Department of Emergency, Disaster and Critical Care Medicine, Hyogo College of Medicine, Nishinomiya, Japan

Shigeo Ohta Department of Biochemistry and Cell Biology, Institute of Development and Aging Sciences, Graduate School of Medicine, Nippon Medical School, Kawasaki, Japan

Liren Qian Department of Hematology, Navy General Hospital, Beijng, China

Jianliang Shen Department of Hematology, Navy General Hospital, Beijng, China

Qiang Sun Department of Hyperbaric Medicine, Navy General Hospital, Beijng, China

Xuejun Sun Department of Navy Aeromedicine, Second Military Medical University, Shanghai, China

Zhouheng Ye Department of Navy Aeromedicine, Second Military Medical University, Shanghai, China

Xiao Zhai Graduate Management Unit, Changhai hospital affiliated to the Second Military Medical University, Shanghai, China

Chapter 1
Hydrogen Element and Hydrogen Gas

Wenwu Liu, Xuejun Sun and Shigeo Ohta

Abstract Features of hydrogen, discovery history of hydrogen, and physical and chemical properties, especially those correlated with the biological effects of hydrogen, are important for the understanding of mechanisms underlying the biological effects of hydrogen. In this chapter, we introduce the formulation, discovery, and physical and chemical properties of hydrogen.

Most substances in the universe are composed of hydrogen element. Even human beings, the most complex form of life on Earth, are composed of hydrogen element. However, few studies have been conducted to focus on the biological effects of hydrogen before the magically biological effects of hydrogen are revealed by scientists in recent years. A secret may still be hidden in these wonder effects of hydrogen.

Though hydrogen has a considerable low solubility in water, its concentration can still approach 0.9 mmol/L. Hydrogen, at this concentration, has been found to be able to confer its protective effects in biological system. Hydrogen is a gas molecule with relatively constant reducibility, an important chemical property of hydrogen. It has been accepted that hydrogen cannot directly interact with substances in biological body and that is why we use hydrogen in diving. However, recent studies reveal that hydrogen is able to neutralize some free radicals conferring protective effects, which maybe a chemical mechanism underlying the biological effects of hydrogen.

Keywords Formulation · Discovery · Physical property · Chemical property

---

S. Ohta (✉)
Department of Biochemistry and Cell Biology, Institute of Development and Aging Sciences, Graduate School of Medicine, Nippon Medical School, Kawasaki, Japan
e-mail: ohta@nms.ac.jp

W. Liu
Department of Diving Medicine, Second Military Medical University, Shanghai, China

X. Sun
Department of Navy Aeromedicine, Second Military Medical University, Shanghai, China

X. Sun et al. (eds.), *Hydrogen Molecular Biology and Medicine,*
DOI 10.1007/978-94-017-9691-0_1

1.1 Hydrogen Is an Element of the Universe

Hydrogen is a chemical element, ranking first in the periodic table with element symbol of "H." It is also the smallest atom in the universe and the simplest element in nature. It has three isotopes, protium, deuterium, and tritium, respectively.

Georges Lemaître is believed to be the smartest scientist alive now in the world, who has proposed the "big bang model" [1]. The idea believes that our universe was originated 15 billion years ago from a pole explosion with infinitely high density, infinitely high temperature, and infinitely small volume. That is the origin of the universe, and also the origin of time and space. The big bang model assumes that, from the moment of the big bang, the universe started to evolve according to Friedman model. The so-called Friedman model refers to the expansion process of the universe, in which, the temperature reduction would happen to any object in the universe because of the radiation (namely heat dissipation) [2]. Temperature is the overall or macroscopic manifestation of average motion velocity of microscopic particles. The faster the particle motion, the higher the macroscopic temperature is, and vice versa. The temperature reduction of the universe would have a huge impact on its state of matter inside. At very high temperature, the particle motion is very fast, and can even escape the limitation of the nuclear or electromagnetic force (at this time no atom would be formed) that attracted them together. It can be expected, when the temperature decreases gradually, the particles attracted by the nuclear or electromagnetic force are connected mutually. More importantly, the kinds of particles in the universe are also dependent on the temperature level.

About 100 s after the big bang, the temperature decreased rapidly from infinitely high to 1 billion degrees. At this temperature, no protons and neutrons have sufficient energy to escape the strong nuclear force between particles that attracted each other, and thus the combination and production of deuterium (heavy hydrogen) nuclei was initiated in the universe. Deuteron consists of a proton and a neutron, and in the evolution process of the universe, the first category of nuclei appeared is the deuterium. Later, deuterons were combined with more protons and neutrons to form helium nucleus, which contains two protons and two neutrons, and also produced a small amount of two kinds of heavier elements, namely lithium and beryllium. The rest of the neutrons decayed into protons and electrons, and proton does not require combination with any other particles, which is the nuclei of protium atom itself, namely the hydrogen ion we are familiar with. In a word, in the formation process of atom, first the deuterium nuclei forms, and then protium nuclei; first the nuclei forms, and then the atom appears.

Four minutes after the big bang, the nucleus production of helium and other elements basically stopped. From a macroscopic consideration, about 300,000 years thereafter, the universe just continued to expand, and nothing happened. Finally, once the temperature decreased to thousands of degrees, the electron and nucleus no longer had sufficient energy to resist the electromagnetic attraction between them, and so they began to combine and form atoms. That is, at 300,000 years after the big bang, with the combination of protons and electrons, hydrogen atom, the greatest and most important member in the universe, was born.

The hydrogen and helium in the galaxies were divided into smaller nebula over time, and they collapsed under their own gravity. When they shrink, the atoms collide inside increasing the gas temperature and finally offer sufficient heat for the thermonuclear fusion reaction initiation. These reactions convert more hydrogen into helium, and the heat released increases the pressure, thus the nebula no longer continue to shrink. Just as the Sun, they would burn hydrogen into helium, and radiate the energy in the form of heat and light. They would remain stable in this state for a long time. The star with bigger mass needs to become hotter in order to balance its stronger gravity, and makes the thermonuclear fusion reaction inside exceedingly faster so that it would consume all of the hydrogen in such a short time, namely 100 million years, and provides light and heat for the life evolution on the planets.

Is our current universe really formed as described above or not? Some scientific observations indicate that in the composition of the universe, 90 % is hydrogen and 9 % is helium. However, the Earth we live does not seem to be in such a ratio. The content of each element of the crust, in descending order, is oxygen, silicon, aluminum, iron, calcium, sodium, potassium, magnesium, potassium, and the percentages of the elements are: 48.06 % of oxygen, 26.30 % of silicon, 7.73 % of aluminum, 4.75 % of ferric, 3.45 % of calcium, 2.74 % of sodium, 2.47 % of potassium, 2.00 % of magnesium, 0.76 % of hydrogen, 0.76 % of others [3]. However, Earth-like planets are not the main component of the universe, but the Sun-like stars with lager mass are members with greater proportion in the universe, which are the major factors determining the universe components.

Hydrogen, as an universal element, can be considered from three aspects. Primarily, hydrogen is the main component in the composition of the universe. Second, the structure of hydrogen is the simplest among all elements in the universe. Third, hydrogen is the core of energy transformation in the universe. As Gerhard Richter, a famous German astrophysicist, said, "Hydrogen is the most important component of the universe." *Legend of hydrogen,* written by Rigdon, a famous science writer and professor in physics at Washington University, was chosen by the magazine *Discover* as the 20 best science books in 2002. Rigdon made comments on hydrogen as follows, "Our understandings of the physical world, no matter to the most basic atom microscopically, or to the universe itself macroscopically, in fact, all can be connected by a hydrogen atom." In *Legend of hydrogen,* the author shows us a unique charm of hydrogen atom in the history of scientific development, and the structure of hydrogen atom is simplest and most unique, and the hydrogen atoms have been challenging the interests and confidences of top scientists globally for centuries. If I have an opportunity to comment on hydrogen, I think hydrogen is born for science, and the hydrogen-centred research is virtually a legendary work in the field of the natural sciences. Moreover, it is now beginning to attract the eyes of biologists, and will continue to reveal its legend in the field of biology.

Coincidentally, the most abundant element in the composition of organic life is hydrogen, and it is an electron transfer media and also a main member of energy transformation in life. Water, the most important component in the milieu interne, is also composed of hydrogen. So, we can say that hydrogen is not only the element of the universe but also the element of life.

While some people say that the world originated from the natural rules, others also believe that it originated from hydrogen. However, in the universe, the Milky Way, the Sun, and the planets, and in the long and exciting origin and evolution process in all kinds of life, hydrogen is everywhere and always plays the most important part.

1.2 The Forms of Hydrogen Element

Hydrogen has three isotopes in the nature world: protium, deuterium, and tritium, and there are also artificially synthesized hydrogen isotopes, such as 4-hydrogen, 5-hydrogen, 6-hydrogen and 7-hydrogen, etc. Protium is the most abundant hydrogen isotope in nature, with an abundance of 99.98 %, and the nucleus of protium has only one proton and no neutrons, and protium is the simplest atom in construct as well (Fig. 1.1) [4]. As the simplest atom, hydrogen has a special value in the theoretical study of atomic physics. Continuous researches into the spectrum, energy level, and bonding of hydrogen by numerous scientists have played a very crucial role in the revolutionary discoveries such as the quantum mechanics.

On the basis of the differences in extranuclear electrons, hydrogen atoms exist in three ways, which are hydrogen ions, hydrogen cations, and hydrogen atoms. In hydrogen ionic compounds, hydrogen atom becomes hydrogen ion (presented as H^-) after gaining an electron to constitute the hydride, and the hydrogen ion is composed of two electrons and one proton, whose volume is the smallest in all known anion except for the electron salt. Hydrogen ion cannot exist in aqueous solution, and it is one of the strongest known bases, and hydrogen ion is also a very strong reductant. Hydrogen atom becomes hydrogen cation after losing an electron (presented as H^+), and in aqueous solution, hydrogen cation exists in the form of hydration.

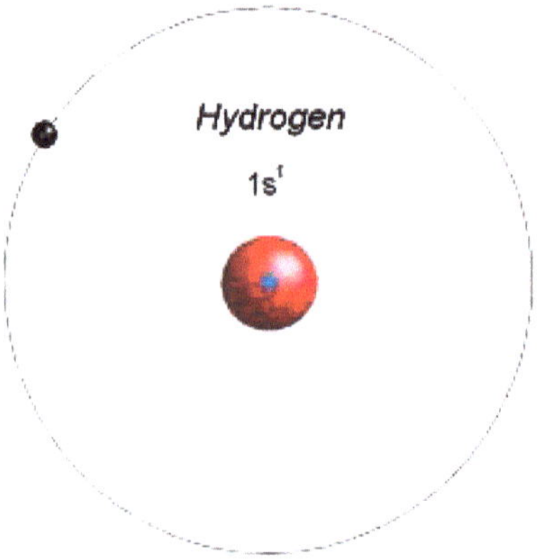

Fig. 1.1 Bohr model of hydrogen atom

Covalent hydrides are compounds formed by nonmetallic, metal-like and some metal (with small electropositivity) elements with hydrogen. Covalent hydride of single nucleus can be presented as MH_{8-x} (x is the valence of M), and some elements may also form a double-nuclear and multinuclear hydrides, such as H_3N and H_2O_2, wherein the carbon-formed hydride accounts for most. Many organic compounds are carbon hydrides, which are the main components constituting the organism structure.

A hydrogen atom and atom X with large electronegative (e.g., fluorine, oxygen, and nitrogen, etc.) are combined with a covalent bond, and when they approach another atom Y with large electronegativity, hydrogen would be a medium between X and Y to generate a bond of X–H…Y, which is called hydrogen bond (X and Y can be the same kind of atom, such as the hydrogen bond between water molecules). Hydrogen bond is an interaction slightly stronger than the intermolecular force and weaker than the covalent bond and ionic bond, and its stability is weaker than that of covalent bond and higher than that of ionic bond. In the case of α-helix of protein, the hydrogen bond is in the form of N–H…O, and in case of double helix of DNA, the hydrogen bond is in the form of NH…O, and NH…N. Because of the prevalence of such hydrogen bonds, these structures are stable. Water and other solvents are heterogeneous, which is also because of the hydrogen bond (O–H—…O) (Fig. 1.2) formed between water molecules. Accordingly, this is also the reason for the formation of hydrophobic binding. Different from covalent bonds, the energy of hydrogen bonds are relatively low, and their formation and destruction are relatively easy, so hydrogen bonds play a very important role in the identification and biochemical reactions between molecules. In organisms, the interactions among hydrogen bonds can have the effect of structure stabilization.

Hydrogen cations are extremely important in acid–base chemistry, and the exchange of hydrogen ions generally exist in acid–base reactions. The so-called pH

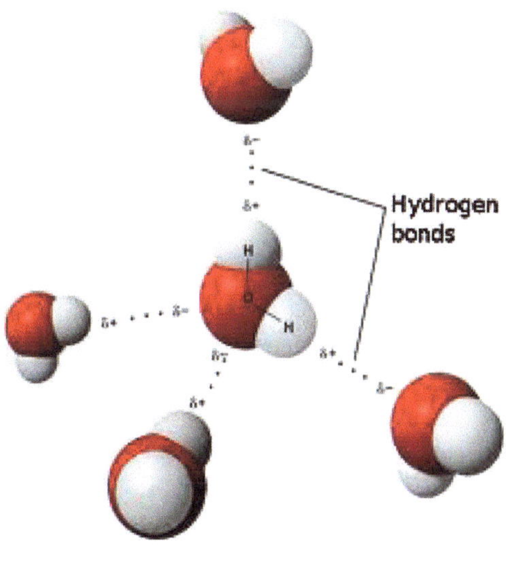

Fig. 1.2 Hydrogen bonds between water molecules

value is also known as the value of acid–base, the concentration index of hydrogen cation or proton, the scaling of hydrogen ion activity in solution, and the measurement standard of acid–base extent in solution as well. Many life activities have very narrow adaptation range of acid–base, for example, the pH of human blood must be maintained between 7.35 and 7.45. Recent biological studies have found that the hydrogen cation itself is a key medium which can activate the channels in cell membranes.

1.3 The History of Hydrogen Discovery

In the history of the discovery of chemical elements, it is difficult to define who discovered the hydrogen, because there were a lot of people engaged in the experiment of hydrogen preparation. In the late sixteenth century, Paracelsus (originally called Theophrastus Bombastus von Hohenheim), a Swiss chemist, noticed a phenomenon that a combustive gas could be generated when the metal was corroded by acid. In another word, he discovered the hydrogen by chance. In 1671, Robert Boyle, a famous Irish philosopher, chemist, physicist, and inventor, also studied the hydrogen, and he made a description about the properties of hydrogen. To whom the scientific discovery belongs is mainly determined by the definition of scientific discoveries itself. In the history of science, the discoverer of hydrogen was ultimately determined to be Henry Cavendish because he was the first scientist who collected the hydrogen, studied hydrogen carefully, and determined some critical properties of hydrogen, such as density [5] (Fig. 1.3).

In 1766, Cavendish submitted a research report entitled "On the artificial air" to the Royal Society of Great Britain. In this paper, in addition to the carbonic acid gases, the main descriptions were about the hydrogen. Cavendish prepared the hydrogen by reacting iron and zinc with the hydrochloric acid and diluted sulfuric acid, and collected the hydrogen by mercury tank method. He found that when a fixed amount of certain metal was reacted with a sufficient amount of each acid, the amount of hydrogen generated is always fixed, regardless of the type and concentration of the acid used. He also found that the mixture of hydrogen and air can be ignited to initiate an explosion. So, Cavendish named the gas as a "flammable air." Furthermore, Cavendish pointed out that this gas is 11 times lighter than ordinary air, and it is insoluble in water or alkali solutions.

In 1781, when Priestley, a British chemist, was doing experiment about "flammable air," he found that some liquid appeared after the explosion of "flammable air" and air mixture. Priestley told his discovery to Cavendish, and Cavendish performed the experiment with different ratios of hydrogen and air, which confirmed the discovery in Priestley's experiment, and determined that the liquid appeared is water. Cavendish pointed out that if the hydrogen and oxygen are put into a glass ball, and then connect with electricity, water would be generated. After the discovery of oxygen, Cavendish repeated the previous experiment using pure oxygen instead of air, and the result not only confirmed that water was generated by the

Fig. 1.3 Cavendish, a famous physicist and chemist in Cambridge University, born in October 10, 1731, in Nice, France, and died in February 24, 1810, in London, Great Britain, at the age of 78. Cavendish is well known for his discovery of the hydrogen and the accurate measurement of the Earth's density

combination of hydrogen and oxygen but also quantitatively determined that about two volume portions of hydrogen and one volume portion of oxygen exactly generate water, and the result was published in 1784. Because Cavendish is a devout supporter of phlogiston theory, so he believed that the metal contains phlogistic elements, when the metal is dissolved in acid, the phlogistic elements contained in the metal is released to form the "flammable air." Although Cavendish is the first person who discovered the hydrogen, and also proved the quantitative relationship between the reaction of hydrogen and oxygen for the first time, because of the shackles of the traditional theories, he did not properly recognize the significance of hydrogen discovery.

Lawrence Lavoisier, a famous French chemist, repeated Cavendish's experiment, and clearly proposed a correct conclusion, that water is not an element, but a compound formed by hydrogen and oxygen. In 1787, Lavoisier confirmed hydrogen was an element, and named it as hydrogen, which means "water-generating element." In 8th May, 1794, Lavoisier died on the guillotine, which was a tremendous disaster in the history of modern science.

There is only a very few free hydrogen on the Earth and in the Earth's atmosphere. Around 1.5 % hydrogen exists in the soil, and oil, gas, animals, and plants also contain hydrogen. In the air of the Earth's surface, the hydrogen accounts 5/10,000,000 (0.5 ppm) of the total volume of air. At the height of 20–25 km from the ground, there are maybe only helium and hydrogen in the atmosphere.

1.4 Physical and Chemical Properties of Hydrogen

1.4.1 Physical Properties of Hydrogen

The common single elemental form of hydrogen is hydrogen gas, and the hydrogen gas is a colorless, tasteless, and odorless diatomic gas molecule. The density of hydrogen is very small, and it possesses the smallest molecular weight in nature, and its density is smaller than air. Under the standard condition (the temperature is 0 °C and the pressure is 101.325 kPa), the mass of 1 l hydrogen gas is 0.089 g. When compared with the same volume of air, the mass of hydrogen is about 1/14 of air. With this property, some people once used hydrogen balloon as a transportation tool [4]. Due to the extremely low density of hydrogen, the hydrogen on the Earth gradually rises into the atmosphere, and finally volatilize into the universe.

Hydrogen is a gas which is very difficult to liquefy, at 101.325 kPa; the hydrogen gas can become colorless liquid at −252.8 °C and the liquid hydrogen has the properties of a superconductor. The liquid hydrogen can be turned into snow-like solid hydrogen at −259.2 °C.

The solubility of hydrogen in common liquids is relatively low. Under certain temperature and pressure, the maximum amount of gas dissolved in a certain quantity of solvent is called the solubility of gas. In addition to the properties of gas and solvent, the solubility of gas is also related to the temperature and pressure. The solubility generally decreases as the temperature increases, due to a huge change in volume of gas that happens when the gas dissolves; the solubility increases significantly with the increase in pressure. Usually, the solubility is represented as the maximum volume dissolved in one volume of solvent at certain fixed temperature. For example, at 20 °C under one atmosphere pressure (pure hydrogen atmosphere), 1.82 ml of hydrogen can be dissolved in 100 ml water, and then the solubility is shown as 1.82 %. If the calculation of solubility is on the basis of molar concentration, the concentration of a pure atmosphere pressure of hydrogen dissolved in water at 20 °C is 0.92 mM. Some studies suggest that, different from many other gases, the solubility of hydrogen may increase as the temperature increases.

As for the solubility of gas dissolved in the liquid, in 1803, Henry, a British chemist, summarized an experience law according to some studies, which was known as Henry's law. According to Henry's law, at a certain fixed temperature and pressure, the solubility of a gas in the liquid is positively proportional to the equilibrium pressure of the gas. In other word, the solubility of gas in the liquid increases proportionally as the partial pressure of the gas increases. Under the same conditions, the solubility of 100 % hydrogen in the liquid is 50 times of 2 % hydrogen. The solubility of hydrogen in water (0.017 %) is slightly higher than that of nitrogen (0.013 %); the solubility of hydrogen in fat (0.036 %) is lower than that of nitrogen (0.067 %), which is approximately 1/2 of nitrogen; at 25 °C) the solubility of hydrogen in ethanol is 0.089 %, which is four times of water. Although the solubility of hydrogen in water and fat is very low, the solubility of hydrogen is very high in metals such as nickel, palladium, and molybdenum; one volume of palladium can dissolve hundreds volume of hydrogen.

Because the molecular weight of hydrogen gas is low, the permeability of hydrogen gas is very strong. It can penetrate rubber and latex tube at room temperature, and can penetrate metal films such as palladium, nickel, steel at high temperature. As for hydrogen balloon which is well filled with hydrogen gas, after placing still overnight, on the next day it cannot fly anymore because of the gas leakage. This is because the hydrogen gas can penetrate through some small invisible pores in the rubber. Moreover, at high temperature and high pressure, hydrogen gas even can pass through extremely thick plates. Due to the strong permeability of hydrogen, when the steel is exposed to the hydrogen gas at certain temperature and pressure, the hydrogen atoms penetrated into the steel lattice and led to embrittlement in the slow deformation. This property of hydrogen gas brings great difficulties for its storage and transportation.

Hydrogen has a large specific heat and excellent thermal conductivity. The thermal conductivity ratio of hydrogen is seven times than that of air. Under the same pressure, the specific heat of hydrogen was 13.6 times than that of nitrogen and 2.72 times than that of helium. Thus, with respect to other gases, hydrogen has a relatively strong endoergic ability and thermal conductivity.

The diffusion rate of hydrogen is fast. According to the gas diffusion law, the diffusion velocity of gas in the liquid is inversely proportional to the square root of molecular weight of the gas. In liquid or human tissue, the diffusion rate of hydrogen is 3.74 times than that of nitrogen and 1.41 times than that of helium (0.138) (see Eq. 1.1):

$$\frac{1}{\sqrt{2}}:\frac{1}{\sqrt{4}}:\frac{1}{\sqrt{28}}=3.74:2.65:1. \quad (1.1)$$

The acoustic speed of hydrogen is fast. Under the standard condition, the acoustic velocity of air is 331 m/s, helium acoustic velocity is 972 m/s, while the acoustic velocity of hydrogen is 1286 m/s. Therefore, if a person breathes hydrogen, the voice will be significantly changed, so the voice of a diver also can be changed after breathing the mixed gas of oxygen and hydrogen.

1.4.2 Chemical Properties of Hydrogen

Hydrogen is chemically stable at room temperature, which is mainly determined by the strong covalent bond between the hydrogen atoms consisting of hydrogen.

Hydrogen is flammable. While igniting or heating, the hydrogen gas can easily react with a variety of substances. When pure hydrogen gas ignites, it can burn quietly, produce a blue flame, give off heat and generate water. If a cold and dry beaker covers on the flame, water droplets can appear on the wall of beaker. The concentration range of hydrogen when the combustion occurs is 4–74 % and the gas cannot burn or explode below or above this concentration range, even under high pressure. In an oxygen atmosphere, the concentration range of hydrogen when the combustion occurs is 4–94 %. When the oxygen concentration is less than 4 %, even under a very high-pressure condition, the mixed gas of hydrogen and oxygen will not burn. The advantage of this property of hydrogen gas has been taken by people in diving operations, and these properties can be used to design safety equipment for breathing hydrogen gas.

Hydrogen has a reductive property. Hydrogen is chemically active, and it can react with oxygen to form water, which is prone to cause combustion and explosion. The flammability of hydrogen is also a manifestation of its reductive property, which is determined by the oxygen-reduction property of hydrogen. Hydrogen can not only react with the elemental oxygen but also react with oxygen in some compounds. For example, when the hydrogen gas passes through hot copper oxide, red copper can be obtained and water is generated. In this reaction, the hydrogen captures the oxygen in the copper oxide to form water; copper oxide loses the oxygen and is reduced to red copper, which proves that the hydrogen gas has a reductive property and is a good reductive agent. Furthermore, hydrogen gas can also reduce some other metal oxides such as tungsten trioxide, ferro or ferric oxide, lead oxide, and zinc oxide.

Although the hydrogen gas has a reductive property, it does not equally mean the hydrogen gas in a solution or organism also has the same property, for example, the concentration of hydrogen should be more than 4 % when oxygen and hydrogen burn, and its ignition point is 400 °C. In the milieu interne of human body, even under the condition of pure hydrogen gas, its dissolution concentration is only 1.8 %, and because the body temperature is only 37 °C, and as that is extremely lower than the condition needed for the reaction of hydrogen and oxygen, the hydrogen cannot react with oxygen in human body. It is exactly an important reason for the hydrogen to be deemed as physiologically inert gas for a long term.

The hydrogen is not only reductive but also oxidative. Hydrogen is a diatomic molecule formed by the covalent hydrogen atoms, and each hydrogen atom can obtain an electron to form a negative hydrogen ion, respectively. This situation is common in the reaction reacted with strongly reductive metal, and its effect is similar to that of chlorine. In this reaction, the hydrogen gas belongs to oxidant and it can oxidize the metal into metal ion. Strictly speaking, the product in the reaction of hydrogen and metal is hydride, and the characteristic of this substance is strongly reductive and very easy to react with water to release plenty of hydrogen.

1.5 A Legend of Nordenau Water

There are legends about the miracle water and longevity water in many regions of the world. In these legends, Nordenau cave in Germany is the most legendary. Japan TV Asahi has a series of program named *To investigate the truth,* which is similar to the column *Approaching Science* in China Central Television's No. 10 station. On 13th June, 1998, Program *To investigate the truth* of TV Asahi first presented an episode of *Panacea: Truth about the miracle water*. And the program reported that the water inside the German Nordenau cave has magical treatment effects to many diseases. The program not only introduced that many patients whose diseases have been cured with the spring but also proposed that the truth of miracle water is the rich hydrogen gas in the water according to the fact that the water contains rich hydrogen gas, which means the root cause of the miracle water is hydrogen gas, and the hydrogen gas has treatment effects to many diseases.

The fame of Lourdes' miracle water in France is much earlier than the miracle water in Germany. A lot of French people are Catholic, but not all disciples go to church to worship, but there is a place that all French people are willing to go, that is Lourdes. Lourdes is a small town located in the southwest corner of Pyrenees, France. It is well known to public for a mystic religious story which describes the miracle water. It was said that in 11th February, 1852, Bernadette, a 14-year-old shepherdess, came to the cave near the bank of Po River to collect some firewood, and suddenly Virgin Mary appeared in front of Bernadette. Virgin Mary told Bernadette: "Please drink some water in the river, and then wash your face!" When Bernadette dug out the ground near the cave, the spring gushed, and later some miracles about different diseases cured by the spring appeared, especially for diseases like paralysis. Just for this reason, this small town has become the biggest Christian pilgrimage, and the number of pilgrims from more than 150 countries is up to 500 million each year, especially for those paralyzed patients on the wheelchair, this place has become the most important holy places for convalescence.

There is a magnificent church built on a hill in Lourdes park, Pyrenees, which is divided into three layers, the upper, middle, and lower layer, in which more than 50 individuals' names were engraved to record the people with long-term disability cured by drinking the miracle water, such as people with lower-limb paralysis were able to stand up and walk after drinking the miracle water. So Lourdes became a religious and cultural tourism city, the containers loaded with the miracle water for selling are available everywhere in the street, and statues of the Virgin emptied for loading the miracle water also are available. Hotels have been equipped with a large number of chain wheelchairs for receiving the disable passengers coming for pilgrim. Nine o'clock in the morning every day, the hotel waiters take these disabled disciples to the church park. The disciples collect the miracle water at the foot of the church hill, and go to the Eye-washing Spring to wash the eyes, then go to the Baptism room and finally listen to the priest's sermon. This saga has become an important part of French histories and cultures.

In 2009, *Lourdes,* the third feature film of Jessica, a famous female Austrian director, was put on screen. This film described the wheelchair-dependent life of Kristen, a woman suffering from paralysis due to multiple sclerosis, and in order to escape the loneness of life caused by the disease, Kristen decided to start a trip to change her fate. Although she was not a Christian, she still decided to go to pilgrimage in Lourdes, the Holy Land of Catholic on the Pyrenees. Magically, she was recovered miraculously there. (Multiple sclerosis is a typical autoimmune disease involving brain and spinal cord, that is an inflammation of brain and spinal cord, and is similar to rheumatoid arthritis. Some exact evidences showed that the holy water in Lourdes, a type of natural spring, contains the highest content of hydrogen gas. Many studies have shown that hydrogen has therapeutic effects on inflammation including rheumatoid arthritis and inflammation of the central nervous system, and maybe Kristen who suffered from multiple sclerosis was really cured by hydrogen.)

Dr. Lin Xiuguang, a famous researcher of electrolysis water, has published many writings about water therapy. *Water for Life,* one of these writings, was a book about the treatment effects of hydrogen-rich water, in which the miracle water with the ability of treating diseases in Nordenau cave in Germany was introduced, in this book, Dr. Lin Xiuguang referenced the article *My tumor become smaller since I visited this cave* published on the magazine *Sauerland journey* in 2002 by Elma, a reporter of the magazine, and many individual cases of patients whose diseases were relieved and cured completely after drinking the miracle water in the cave, were also included in the article. What has to be emphasized was that, although these cases are very important, it cannot be confirmed that these diseases can be treated with water or hydrogen. In medical science, many standards have to be considered before a therapy or drug enters into clinical applications, and the most commonly used standard is a double-blind randomized controlled trial. In a proper way, these cases do not meet the basic clinical research standards, so the credibility is limited. However, these cases provide our medical researches with some very important clues, and based on these clues we can perform further investigations. Therefore, it is necessary to sort out and summarize these cases.

Varta, a person from Sindelfingen, Germany, and his arthritis and cataracts have been cured over 4 treatments. Lizzy, a person from Clausthal, and her gallstones have been completely cured. Furthermore, there are more cases about the relief and curative effects of patients' diseases such as diabetes, rheumatoid arthritis, eczema, depression, atherosclerosis, insomnia, nephritis, myocardial infarction, bronchitis, headaches, brain tumors, late breast cancer, sequelae of viral pneumonia, epilepsy, and deafness.

The academia has called the treatment of diseases by Nordenau water as *Nordenau phenomenon.* In 2006, the Fifteenth Japan Society of Animal Cell Technology conference, a collaboration research from Germany and Japan researchers, about the effect observation of the German Nordenau cave water on the treatment of type II diabetes patients, summarized 411 cases of type II diabetes patients, with an average mean age of 71-years-old, and the subjects drink 2 l of Nordenau cave water every day for 6 days in average. According to the results of glucose, lipids, and inosine before and after drinking spring water, the researchers found that drinking Nordenau water has a relatively good effect on diabetes.

It must be clear that the legendary curative effects of the holy water or miracle water mentioned above absolutely cannot be deemed as a final conclusion. Because these results are very likely to be affected by the user's psychological implications, and the treatment effects of these diseases by holy water or miracle water largely come from individual cases, so it is impossible to directly conclude that this kind of disease can be cured definitely by hydrogen gas, because of lack of rigorous and standard basis of scientific researches. But now, many animal experiments and some clinical studies on the biology of hydrogen have demonstrated that the hydrogen gas may have potential values on the treatment of diseases, such as diabetes, rheumatoid arthritis, eczema, depression, and atherosclerosis. As for the reason why the hydrogen gas may have effects on the treatment of these diseases, it will be introduced systematically in the following chapters of this book.

1.6 Hydrogen Gas Related Products

The program producers and Asahi TV probably did not expect that the report on the miracle water in Germany would unexpectedly turn out to be an important reason for promoting the development of hydrogen-related products by the health industries in Japan. After watching this program, some Japanese businessmen thought that, since the effect of this miracle water is because of the role of hydrogen, whether the miracle water with same effect can be produced by adding the hydrogen gas to the ordinary water. As for this statement, I have heard from technical personnel responsible for the development of hydrogen-related products repeatedly. In this sense, the program producer has made important contribution to the study on the biological effects of hydrogen.

Among the products developed by Japanese companies, there are some relatively representative products such as negative hydrogen ion, magnesium hydrogen-water stick, saturated hydrogen water, hydrogen-containing cosmetics. All these products were manufactured under the premise that the hydrogen has an antioxidant effect, and presume that the hydrogen can play a role in the prevention and treatment for a variety of oxidative damage. Of course, as healthcare products and daily necessities, the manufacturers cannot directly promote the effect on treatment of disease, but it is indeed the intrinsic value and selling point of these products.

Wherein, the hydrogen-water stick is to use the principle that the metal magnesium can react with water to produce the hydrogen gas, and meanwhile a small amount of magnesium ions is generated, which also belongs to the trace element that is beneficial. To make the taste much better, these products generally would be added with some tourmaline materials. In order to extend the lifespan of the hydrogen-water stick, it is necessary to process the metal magnesium such as calcination, so as to slow its reaction with water and stabilize it. These methods are some relatively conventional techniques in the field of materials.

The basic principle of saturated hydrogen water is to directly dissolve the hydrogen gas in water. In our common sense, the hydrogen gas is not soluble in water, in the chemistry experiment on hydrogen production of high school, we once used the water draining method to collect the hydrogen gas, and the main reason is due to the water-insoluble characteristics of hydrogen. In fact, it does not mean that the hydrogen gas cannot dissolve in water, but actually it is of relatively low solubility. For example, at 20 °C the solubility of hydrogen in water is 1.5 ml/100 ml. It means that, 1 l of water can dissolve 15 ml of hydrogen. If it is converted into the mass, there is about 1.34 mg of hydrogen. Whether the hydrogen dissolved in liquid can achieve the antioxidative effects (because it has a low solubility) will be analyzed in detail in the following chapters of this book.

The biggest obstacle encountered in the development of saturated hydrogen water is not the low solubility of the hydrogen, but that the hydrogen is very easy to release from water because the hydrogen molecule is very small and thus it is very easy to escape out of the packing material and tiny crevices. Therefore, the requirements on packaging materials are relatively high. At present, Japanese companies commonly use aluminum packaging materials, which can extend the storage time of hydrogen to 6 months or even more than a year. In the study of the biological effects of hydrogen, the saturated hydrogen water has relatively important contribution. Our research group finished many important studies on the biological effects of hydrogen, just because we are the first in preparing the saturated hydrogen saline for injection in the world, and then these important studies were achieved on the basis of the extensive cooperation performed with more than 50 domestic medical universities and general hospitals. Institute of Gerontology in Nippon Medical School is the first research unit studying the hydrogen effects on the treatment of diseases, funded by Blue Mercury, a leading Japanese manufacturer of hydrogen water, and the first international research center of molecular hydrogen medicine was established in Nippon Medical School. As for the current researches in Japan, most of the saturated hydrogen water used is supplied directly by these companies. In a sense, the study of hydrogen biology was started from the enterprises and gradually attracted attentions in academic field.

1.7 Outline of Studies on Biological Effects of Hydrogen

At the beginning of this century, while the business community was developing a wide range of hydrogen-related products, the biological effects of hydrogen has also been paid attention by the academia. There were nearly ten research groups involved in the study on the effect of hydrogen, and the Institute of Gerontology in Nippon Medical School is the earliest one to explore the biological effects of hydrogen and also the most successful one. Shigeo Ohta, the chief professor in the Institute of Gerontology in Nippon Medical School, is a senior expert of geriatric research, and his main research area in the past was the relationship between mitochondria and aging. He is also the president of Japan Mitochondria Society, Society of Cell Death and Society of Hydrogen Biology Research, and he has very high academic position in the international biochemistry academia, especially in Asia. Commissioned by some Japanese hydrogen companies and led by Professor Shigeo Ohta, the Institute of Gerontology in Nippon Medical School began the hydrogen-related research in the treatment of diseases since 2005. In July 2007, after nearly 4 years of painstaking research, they finally demonstrated that if people breathe a small amount of hydrogen (1–4 %, breathe for 35 min), the hydrogen breathed would have strong selective antioxidant effect, and through neutralizing the toxic free radical, the hydrogen significantly reduced the cerebral infarction caused by cerebral ischemic and reperfusion damage (Fig. 1.4) [6]. The treatment effect of hydrogen on the cerebral ischemia was similar to that of the immunosuppressant

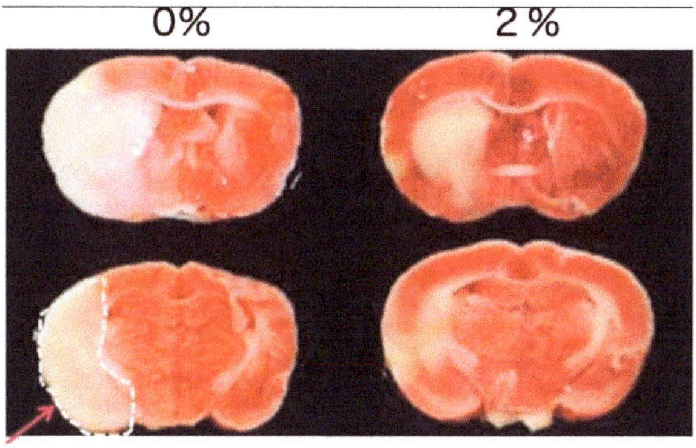

Fig. 1.4 Typical change of cerebral ischemia treated by hydrogen, TTC staining can clearly distinguish between normal and damaged brain tissue, TTC staining of normal brain tissue is red, TTC staining of brain tissue with damage or infarction is white. In the figure above, the *left ones* are the brain tissues with unilateral ischemia in rats which did not breathe the hydrogen gas, and the *right ones* are the brain tissues with unilateral ischemia in rats which breathed 2 % hydrogen gas. The results show that the treatment effect of hydrogen is very significant. TTC 2,3,5-triphenyltetrazolium chloride. (Permission obtained from [6])

FK-506, and was significantly better than edaravone, an antioxidant drug currently used in the clinical treatment of cerebral ischemia. Their research papers were published as an original article in *Nature Medicine,* a world-famous journal. This study not only led to a huge social impact and global media reports (Fig. 1.5) but also generated a sensation in the academia. Then they reported other studies about treatment of liver and cardiac ischemia with breathing a small amount of hydrogen gas. These study reports published by the Japanese scientists completely changed the understanding of hydrogen in the academia, and quickly attracted wide attentions among scientists in Japan, USA, and China, and 5 years later, the paper published in *Nature Medicine* had been cited for more than 300 times, and became a classic paper in this academic field.

Our research group has long been engaged in the study on diving hyperbaric medicine, while the main research objective of diving hyperbaric medicine is the various types of gas which can be breathed, and the hydrogen is just the type of gas which is the research focus in the field of diving hyperbaric medicine. In diving hyperbaric medicine, it has long been believed that hydrogen is the same as nitrogen and helium, and all of these are physiologically inert gas. The so-called physiologically inert gas, is the gas that can be breathed by the human body but does not chemically react with any substance, and would exclude all the gas inhaled completely through respiration. As an important international team in the field of diving medicine, we have long been convinced that the biological effects of hydrogen is already very clear, namely that it has no obvious biological effect, but only has a few effect at extremely high dose. The biological effects of hydrogen discovered by the Japanese scientists revolutionized the traditional view in the academia, which made us extremely shocked. Except for the shocking, definitely we were

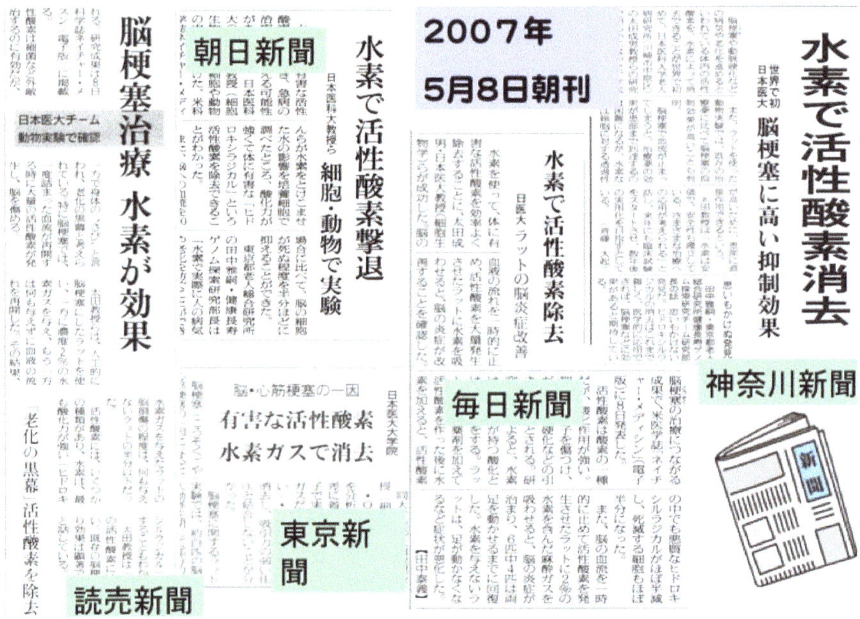

Fig. 1.5 Major media reports about the research of hydrogen biology in Japan

very interested in it, and we learnt and analyzed these documents with a suspicious mind. Our understanding at that time was that the effect of hydrogen was so surprising, since the study found that only breathing with 2 % hydrogen for 35 min, it can be very effective in the treatment of cerebral ischemic and reperfusion damage. But after going through this article at that time, we just were in the surprising and did not show much more interest, and even were still skeptical. And then another 2 months later, another research report was published by Nippon Medical School in another very famous journal *Biochem Biophys Res Commun,* which reported that the hepatic, ischemic, and reperfusion damage can be treated by breathing hydrogen [7]. This study was very simple and only had three aspects, wherein the first one used the conventional pathology staining to prove hydrogen can protect the liver tissue from damage, and another one used a very simple indicator, MDA, which can reflect the oxidative damage of tissue, to prove hydrogen has antioxidant effect. The third method was to detect the liver function, namely the transaminase activity we commonly used. The three research methods or indicators, it is simply effortless for many common life science research laboratories. At that time, we thought that these Japanese scientists were simply picking up the papers, and it was so easy and so convenient. As a researcher, of course, we could not stand the temptation of discovering this new phenomenon, so we began to prepare to do this work and to pick up some papers by the way. Therefore, we began the study of hydrogen on the treatment of diseases, at the beginning we used similar means applied by the Japanese scientists to prove that breathing hydrogen can treat the hypoxia-ischemic encephalopathy in neonatal brain, we only spent a very short time before completing the research, and published it quickly on *Neuroscience Letters* [8]. Generally speaking, the number of citation is an important international standard on the evaluation of paper, although *Neuroscience Letters* is not the peer journal globally, but 5 years later, this paper has been cited for more than 100 times, and it ranks one of top ten papers in more than 5000 papers published in the journal in 5 years, which should be a quite good achievement.

In 2008, Nippon Medical School published a paper once again that proved that the hydrogen breathing can treat the myocardial ischemia, which is obviously understandable, to study the cerebral ischemia and hepatic ischemia initially, and then to study the myocardial ischemia, and more studies should be started in the future, which also indicated that the group's highlighting and affirming on the issue [9]. At this time, a study about the hydrogen breathing on the treatment of intestinal transplantation from University of Pittsburgh, a famous university from USA, had been published too [10]. University of Pittsburgh represents the highest level of organ transplantation among the academic institutions worldwide, in which, the first transplantation cases of human liver, heart, and kidney had been finished. Various types of organ transplantation have achieved very good results now, however, because the lung and intestine are prone to be inflammatory during the transplantation, which has long been an obstacle hard to be conquered in the academia. One team from the university had performed a study on the use of hydrogen against the intestinal damage in organ transplantation. From the pathological section provided in the article, it is very clearly found that the treatment effect of hydrogen is very significant (Fig. 1.6), and they also have proved that the hydrogen has anti-inflammatory effect for the first time globally. Because hydrogen has anti-inflammatory effect, it indicates that hydrogen may have therapeutic effects on other types of inflammation; this research has made an important impact on the studies on the biological effects of hydrogen. Five years later, the paper has been cited for 120 times, which has become the most influenced paper except for the papers published by Nippon Medical School. After that, University of Pittsburgh continuously reported some systematic studies on the treatment effects of hydrogen on transportation of heart, lung, kidneys, and vascular systems.

Why the studies on hydrogen biology quickly generate widespread attention in the academia? Why hydrogen can produce the treatment effects on diseases? The explanation given by Nippon Medical School in their paper published in 2007 is that the hydrogen has selective oxidation. Here, we will briefly explain what selective antioxidation of hydrogen is.

Free radicals are atoms, atom groups, or molecules containing unpaired electrons. Free radicals are necessary substances maintaining the normal life, and the radical reaction is one of the most basic types of biochemical reactions, and is the basis of energy metabolism in organism. There are many types of free radicals, and part of free radicals is an important intracellular signaling molecule, which have biological function and positive effect. Another part of free radicals is biological macromolecule, a dangerous killer of cells. Under physiological conditions, free radical is generated continuously and also continues to be cleared so that it is maintained at a normal physiological level; too much or too little free radical can

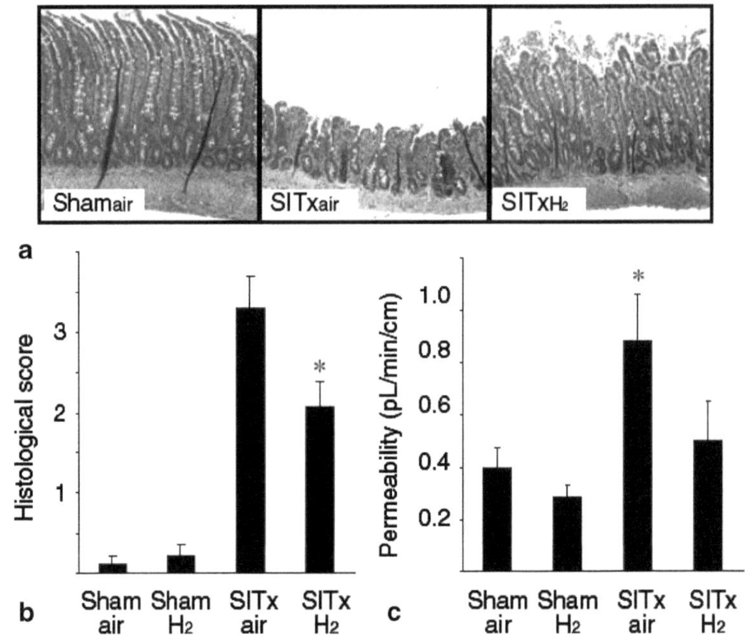

Fig. 1.6 The protective effects of hydrogen on intestinal damage after transplantation. **a** HE staining of tissues, from *left* to *right* is normal intestinal tissue, intestinal tissue after transplantation, and intestinal tissue with treatment of hydrogen after transplantation, in which it can clearly distinguish that the treatment of hydrogen can make a significant improvement on the intestinal tissue with damage. **b** The quantitative analysis data on the basis of Figure a. **c** Using the permeability change as the index of intestinal function, the results indicate that the treatment of hydrogen can reduce the intestinal permeability and increase the intestinal function. *HE* hematoxylin and eosin stain. (Permission obtained from [10])

adversely affect and even damage the body, respectively. The type of free radicals in an organism includes semi-quinone, oxygen, carbon, and nitrogen radicals, etc., in which the majority of researches are on oxygen and nitrogen free radicals. Oxygen radicals contain superoxide anion, singlet oxygen, and hydroxyl radicals, and due to hydrogen peroxide is similar oxygen free radical with respect to the biological effects; generally oxygen free radical, and hydrogen peroxide are called active oxygen together. The more important nitrogen radicals contain nitric oxide and peroxynitrite anion [11].

When the organs or tissues in an organism appear ischemic or with inflammation, a large number of various types of active oxygen will be generated in the cell, among these active oxygen, hydrogen peroxide and nitric oxide have very important signal effect, and their toxicities are weak, but the toxicities of hydroxyl radicals and peroxynitrite anion are strong, which are the main medium leading to cell oxidative damage. In the past, the research idea with respect to treatment of oxidative damage is to find out reductive substance strong enough, but too strong reductant will inevitably lead to an imbalance of endogenous redox condition, and even is the main reason leading to the ineffectiveness of antioxidant therapy. Therefore, to find out the substance which can selectively neutralize the hydroxyl radical and peroxynitrite anion, is an effective method for the treatment of various types of oxidative damage, and the right idea which should be chosen for antioxidation.

Currently, progress on the discovery and study of selective antioxidants is still relatively slow, and there is relatively few known selective antioxidant. In 2007, a paper published by the Japanese scientists has clear evidences to prove that hydrogen has selective antioxidative effect, in another word, hydrogen is a selective antioxidant. Whether hydrogen is the best ideal selective antioxidant or not, more researches still are needed to support the opinion.

As a selective antioxidant, hydrogen has a very attractive prospect in the treatment of diseases, but breathing hydrogen is not a very satisfactory mean because the mixture of hydrogen and oxygen can lead to combustion and explosion, there is a certain danger, so are there some better means and methods for using hydrogen? In fact, the principle of breathing 2% hydrogen is that hydrogen is dissolved in the blood and act after reaching the organs such as brain through circulation. The amount of gas dissolved is related to concentration; the dissolution concentration of 100% hydrogen is more than 60 times of that of 2% hydrogen dissolved. If we dissolve 100% hydrogen in water, then we can use a small amount of hydrogen solution to play the same role. This hydrogen solution is a hydrogen-rich water or hydrogen-rich saturated water, generally is abbreviated to hydrogen water. In the early stage of research on hydrogen effect in 2008, we have encountered the problems of these theories; as we have long been engaging in the diving gas medical research, all the knowledge belong to our expertise, so it is easy to understand. So, we have prepared the saturated hydrogen saline for the first time worldwide and carried out a series of studies.

First, we used the animal disease model same as in the experiment of hydrogen breathing, to prove that injection with a small amount of this saline (similar to 500 ml in the injection of adult) and can be very effective on the treatment of cerebral hypoxic-ischemic encephalopathy in neonates. Because this model used 8-day-old animals, it could not perform more complex behavioral research; we fed the test and control animals for 1 month, and then water maze swimming experiment was used to test the animal's learning and memory ability, and it was found that the learning and memory ability of untreated animals was significantly decreased, while the learning and memory ability of treatment group was significantly increased.

We have cooperated with some units such as Shanghai Changhai Hospital, to perform researches on the intestinal damage after ischemia in the animal, and soon we found this saline to have an obvious therapeutic effect on the decrease of intestinal motility, intestinal tissue apoptosis, and lung inflammation induced by the intestinal ischemia (Fig. 1.7) [12]. We have cooperated with Shanghai Changzheng Hospital, to prove that saline injection can be used for the treatment of acute pancreatitis [13]. The most effective treatment medicine for acute pancreatitis is protease inhibitor, although the effect is clear, but the price is very expensive. If the hydrogen gas can be effective in treatment of diseases, even if it can reduce the usage amount of

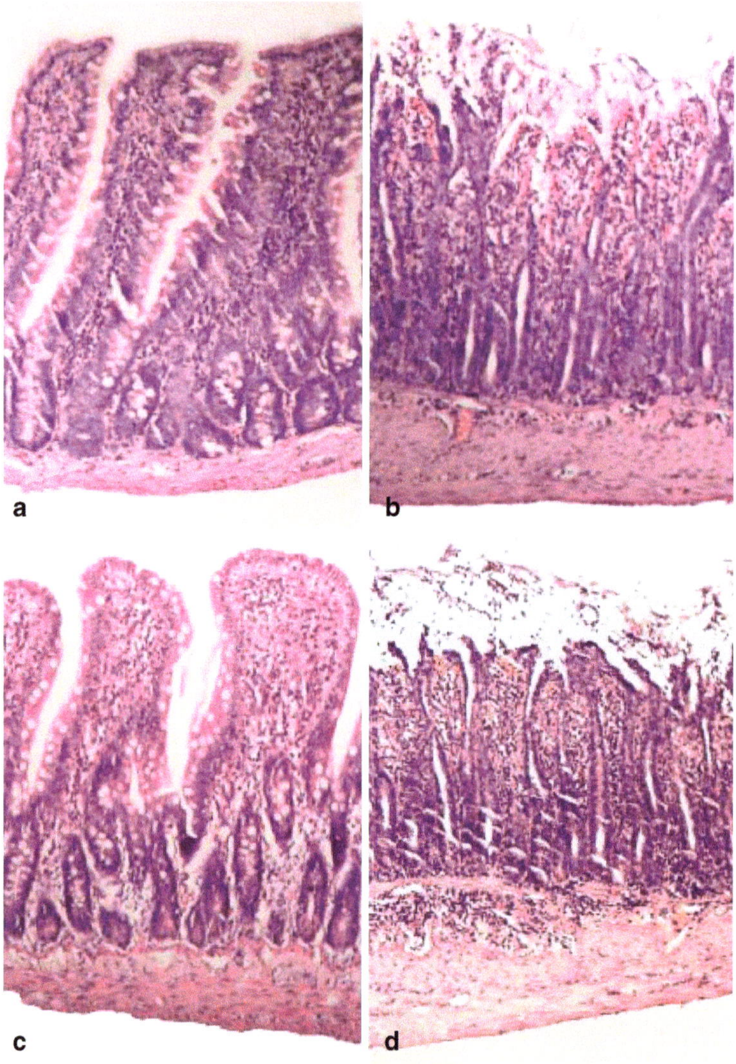

Fig. 1.7 The therapeutic effect of hydrogen saline on intestinal ischemic and reperfusion damage. **a** The normal control group; **b** The intestinal ischemia, **c** the intestinal ischemia group with hydrogen treatment; **d** Intestinal ischemia control group with nitrogen treatment. (Permission obtained from [12])

protease inhibitor or can enhance the therapeutic effect, which clearly can save a lot of treatment cost. Spinal cord trauma can lead to paralysis and seriously affect the quality of patient's life. Currently, there is no very effective treatment method, and our research found that the use of hydrogen saline in the early spinal cord trauma, can recover the movement capacity of animal with hind limb paralysis after spinal cord damage, and reduce the number of necrotic nerve cells at the damage site [14].

Pancreatic head tumor can cause ampulla oppression and lead to bile duct obstruction and intrahepatic cholestasis, and then generate liver damage, cirrhosis, ascites, and weight loss. We have cooperated with Eastern Hepatobiliary Surgery Hospital and found that if hydrogen saline is injected intraperitoneally continuously after bile duct obstruction, it can reduce liver damage and ascites, and significantly increase the body weight [15]. Long-term smoking can cause chronic obstructive pulmonary damage, and we have cooperated with the Department of Physiology, Medical College of Fudan University, and found that the use of animal model of chronic smoking to prove that hydrogen can reverse the pulmonary hemorrhage and inflammation in animal in 1 week. We have cooperated with the Department of Neurology, Affiliated Hospital of China Medical University, to carry out the study on senile dementia in animal, and found that hydrogen can effectively treat the senile dementia in animal, and its effects are closely related to some important signaling pathways influencing the necrosis of neuron. We have cooperated with Professor Qin Shucun, Taishan Scholar from the Institute of Atherosclerosis in Taishan Medical College, to conduct the research on hydrogen treatment of atherosclerosis. The study found that injecting hydrogen saline can effectively treat the atherosclerosis caused by high-fat diet and genetic defects [16]. We also found that injection with hydrogen saline can treat diabetes and diabetic retinal damages [17].

We have heard a true folktale during that period occasionally; it was said that a few decades ago, a famous Shandong fertilizer factory found a strange phenomenon that the intelligence and movement capacity of many patients with carbon monoxide poisoning were significantly recovered if only they worked in the hydrogen purification workshop for many years, while patients in other positions were not so lucky. Inspired by this legend, we have carried out the study of hydrogen on the treatment of delayed encephalopathy caused by carbon monoxide poisoning, and it was found that hydrogen saline can significantly relieve after the change of cerebral nerve demyelination after carbon monoxide poisoning, meanwhile the inflammation and apoptosis were significantly decreased, and the paper was published on *Emergency Medicine,* a famous international journal. We also found that hydrogen has preventive and therapeutic effects on driving decompression and lung damage caused by chronic oxygen toxicity. Since 2008, we have published 50 papers in the field of treatment of various diseases with hydrogen, many hydrogen effects in these studies were first discovered in the world, the number of our research papers were more than the sum of the second and third research teams currently ranked in this field. We have become the team with maximum papers published in this field worldwide, and the number of citations and the grade of journal published were similar to that of international counterparts. China also has become the country with maximum papers in the field internationally because of our huge number of studies published.

In recent years, some Japanese scientists have conducted research on therapeutic effect of drinking hydrogen water on the diseases at the same time. They found that the hydrogen water has a therapeutic effect on patients with diabetes, and has a preventive effect on atherosclerosis, has a therapeutic effect on Parkinson's disease, and has a good therapeutic effect on fatty liver and obesity (Fig. 1.8).

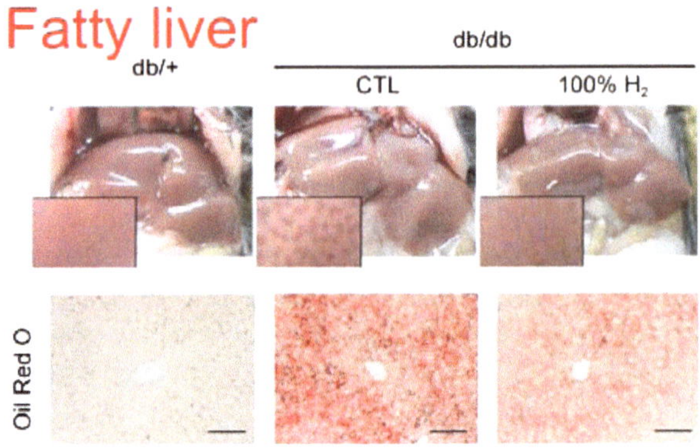

Fig. 1.8 Therapeutic effect of hydrogen water on the fatty liver in mice. (Permission obtained from Kamimura et al. Molecular hydrogen improves obesity and diabetes by inducing hepatic FGF21 and stimulating energy metabolism in db/db mice. Obesity (Silver Spring). 2011;19(7):1396–403)

Until 2012, hydrogen medical research has become an international hotspot, and there are more than 300 academic papers published about the treatment effects of hydrogen. These papers involve various types of organ and tissue ischemia, atherosclerosis, diabetes in human, organ and systemic inflammation, trauma, senile dementia, and other degenerative diseases of central nervous system, depression, arthritis, and radiation damage [18].

The discovery that hydrogen can treat a disease, not only provides us with a prospect of safe and effective treatment of disease but also has a profound impact on reevaluating the life phenomenon of oxidation and antioxidation for us. In subsequent chapters of this book, we will not only analyze in depth on the treatment mechanism of disease treated by hydrogen but also summarize and analyze the related history and current status of oxidation and antioxidation to arouse the readers to think deeper about the oxidation and antioxidation.

When we recall the research history of the biological effects of hydrogen, some thoughts often come into our mind. If there was no study performed by the Japanese scientists, we may never pay attention to these studies on hydrogen. Because we had more understandings of biological effects of hydrogen in the past, and under the conditions with relatively rough detection indicators, it was hard to obtain any association that hydrogen has a therapeutic effect on the diseases. If the Japanese scientists also had relatively good understandings of this knowledge, we are not sure whether they would carry out such a similar study. Therefore, sometimes the old knowledge may adversely affect the academic innovation, but many significant discoveries are just from some unexpected, even ridiculous studies.

References

1. McCrea WH. A philosophy for big-bang cosmology. Nature. 1970;228(5266):21–4.
2. Friedman A. On the curvature of space. Gen Relat Gravit. 1999;31(12):1991–2000.
3. Israel science and technology homepage. List of periodic table elements sorted by abundance in Earth's crust. Israel science and technology homepage. http://www.science.co.il/PTelements-Hebrew.asp. Accessed 15 Apr 2007.

4. Staff. Hydrogen (H2) properties, uses, applications: hydrogen gas and liquid Hydrogen. Universal Industrial Gases, Inc. 2003. http://green.wikia.com/wiki/Hydrogen. Accessed 5 Feb 2008.
5. http://en.wikipedia.org/wiki/Hydrogen
6. Ohsawa I, Ishikawa M, Takahashi K, Watanabe M, Nishimaki K, Yamagata K, Katsura K, Katayama Y, Asoh S, Ohta S. Hydrogen acts as a therapeutic antioxidant by selectively reducing cytotoxic oxygen radicals. Nat Med. 2007;13(6):688–94.
7. Fukuda K, Asoh S, Ishikawa M, Yamamoto Y, Ohsawa I, Ohta S. Inhalation of hydrogen gas suppresses hepatic injury caused by ischemia/reperfusion through reducing oxidative stress. Biochem Biophys Res Commun. 2007;361(3):670–4.
8. Cai J1, Kang Z, Liu WW, Luo X, Qiang S, Zhang JH, Ohta S, Sun X, Xu W, Tao H, Li R. Hydrogen therapy reduces apoptosis in neonatal hypoxia-ischemia rat model. Neurosci Lett. 2008;441(2):167–72.
9. Hayashida K1, Sano M, Ohsawa I, Shinmura K, Tamaki K, Kimura K, Endo J, Katayama T, Kawamura A, Kohsaka S, Makino S, Ohta S, Ogawa S, Fukuda K. Inhalation of hydrogen gas reduces infarct size in the rat model of myocardial ischemia-reperfusion injury. Biochem Biophys Res Commun. 2008;373(1):30–5.
10. Buchholz BM1, Kaczorowski DJ, Sugimoto R, Yang R, Wang Y, Billiar TR, McCurry KR, Bauer AJ, Nakao A. Hydrogen inhalation ameliorates oxidative stress in transplantation induced intestinal graft injury. Am J Transplant. 2008;8(10):2015–24.
11. Dröge W. Free radicals in the physiological control of cell function. Physiol Rev. 2002;82(1):47–95.
12. Zheng XF, Mao YF, Cai JM, Li YH, Liu WW, Sun PL, Zhang JH, Sun XJ, Yuan HB. Hydrogen-rich saline protects against intestinal ischemia/reperfusion injury in rats. Free Radical Res. 2009;43(5):478–84.
13. Chen H1, Sun YP, Li Y, Liu WW, Xiang HG, Fan LY, Sun Q, Xu XY, Cai JM, Ruan CP, Su N, Yan RL, Sun XJ, Wang Q. Hydrogen-rich saline ameliorates the severity of l-arginine-induced acute pancreatitis in rats. Biochem Biophys Res Commun. 2010;393(2):308–13.
14. Chen C, Chen Q, Mao Y, Xu S, Xia C, Shi X, Zhang JH, Yuan H, Sun X. Hydrogen-rich saline protects against spinal cord injury in rats. Neurochem Res. 2010;35(7):1111–8.
15. Liu Q, Shen WF, Sun HY, Fan DF, Nakao A, Cai JM, Yan G, Zhou WP, Shen RX, Yang JM, Sun XJ. Hydrogen-rich saline protects against liver injury in rats with obstructive jaundice. Liver Int. 2010;30(7):958–68.
16. Song G, Tian H, Qin S, Sun X, Yao S, Zong C, Luo Y, Liu J, Yu Y, Sang H, Wang X. Hydrogen decreases athero-susceptibility in apolipoprotein B-containing lipoproteins and aorta of apolipoprotein E knockout mice. Atherosclerosis. 2012;221(1):55–65.
17. Xiao X, Cai J, Xu J, Wang R, Cai J, Liu Y, Xu W, Sun X, Li R. Protective effects of hydrogen saline on diabetic retinopathy in a streptozotocin-induced diabetic rat model. J Ocul Pharmacol Ther. 2012;28(1):76–82.
18. Ohta S. Molecular hydrogen as a preventive and therapeutic medical gas: initiation, development and potential of hydrogen medicine. Pharmacol Ther. 2014;144(1):1–11.

Chapter 2
Absorption and Release of Hydrogen Gas in Body

Dong Cao, Zhouheng Ye and Wenwu Liu

Abstract The definition of inert gases is different in the fields of chemistry and physiology. Physiologically, inert gases mainly include hydrogen, helium, and nitrogen and refer to those that cannot react with other substances in human body although hydrogen is a highly active gas in chemistry. Under normal condition, the human body is saturated by nitrogen. When we inhale another inert gas at a high pressure or normal pressure, the new inert gas may enter the human body in the drive of pressure gradient force. The law of saturation and desaturation of inert gases has been summarized by Haldane, a Scottish physiologist. In this chapter, we discuss the saturation and desaturation of inert gases, with nitrogen as an example. Drinking hydrogen water or injection with hydrogen saline has similar pattern in the absorption and washout of hydrogen in human body to the inhalation of hydrogen except for the high velocity for hydrogen. Also, since hydrogen diffuse rapidly, it could release through skin when the concentration of hydrogen is out of capacity.

Keywords Diffusion · Blood · Saturation · Desaturation · Concentration

2.1 Hydrogen Gas Is Physiologically Inert Gas

Inert gases that are often mentioned in diving medicine have different concepts and meanings from the chemically inert gases. The chemically inert gases refer to those molecules with the outermost shell of their atom filled with electrons, and mainly consist of helium family gases, namely helium, neon, argon, krypton, xenon,

W. Liu (✉)
Department of Diving Medicine, Second Military Medical University, Shanghai, China
e-mail: liuwenwu1980@hotmail.com

D. Cao
Department of Osteology, The sixth people's hospital, Shanghai, China

Z. Ye
Department of Navy Aeromedicine, Second Military Medical University, Shanghai, China

X. Sun et al. (eds.), *Hydrogen Molecular Biology and Medicine*,
DOI 10.1007/978-94-017-9691-0_2

Table 2.1 Commonly used physical parameters of various inert gases

Gas	MW	Density (g/L, STP)	Ratio (with air as 1)	Viscosity (μPa·s, STP)	Diffusion coefficient (cm^2/s, STP)	
					In water	In air
H_2	2	0.09	0.0695	8.4	[a]5.2×10^{-5}	0.63
He	4	0.18	0.138	18.6	7.9×10^{-5}	0.503
Ne	21	0.90	0.695	29.8	3.48×10^{-5}	0.222
N_2	28	1.25	0.967	16.6	3.01×10^{-5}	0.190
Ar	40	1.79	1.379	21.0	2.52×10^{-5}	0.159
Kr	83.8	3.70	2.868	23.3	1.75×10^{-5}	0.110
Xe	131.5	5.58	4.525	21.0	1.39×10^{-5}	0.088

STP standard conditions (temperature, pressure), *MW* molecular weight
[a] Measured at 21°C

and radon. These gases have very good chemical stability and are difficult to participate in chemical reaction under general condition; therefore, they are termed as inert gases. However, inert gases in diving medicine and physiology, which are also known as "neutral gases," refer to those gases only present within the body in a physically dissolved state, and maintain their original nature. They do not react with other substances in the body, do not participate in the body's metabolism, and freely diffuse according to the partial pressure gradient of the gases in vitro and in vivo. By definition, helium, nitrogen, and hydrogen are used as inert gases in diving medicine (Table 2.1). However, according to the results in hydrogen molecular biology, hydrogen may have some reactions with other substances in vivo. Then whether gases that can react with other substances are still considered inert gases? The only effective means to solve this problem is to modify the definition of inert gas in diving medicine. From this perspective, the research on the biological effects of hydrogen can at least lead to a reconsideration about the gases in diving medicine, and also have a very important role in promoting studies in diving medicine.

The inert gas that is most commonly used in medicine and physiology is nitrogen. But its outermost shell is not filled with electrons, so it does not belong to chemically inert gases; hydrogen has an active chemical property, but it has been demonstrated as an inert gas used in diving. Helium is an inert gas often used in deep diving. Thus, it is obvious that inert gases in diving medicine-physiology not only include some chemically inert gases but also include some nonchemically inert gases, or even chemically active gases. Therefore, these gases should be interpreted as "physiologically inert gases."

Inert gases are not useless or dispensable for human survival; instead, they are indispensable component of the gas medium to maintain life. To maintain the oxidation–phosphorylation energy metabolism which is essential to life, the inhaled gas must contain a certain proportion of oxygen. But if the partial pressure of oxygen is too high or even pure oxygen, then it will cause damage to the body or even death since the presence of excessive oxygen is toxic to the body. Typically, the air we breathe contains 21 % of oxygen. Oxygen of this concentration and pressure, which is termed as normoxia, is optimal for human survival. It is the 78 % of nitrogen contained in the air that helps maintain the state of normoxia. Nitrogen plays a role of "oxygen diluent" in this condition. When doing diving/high-pressure operations, pure oxygen is only used in some certain forms of shallow depth or short duration, while inert gases shall be used in other forms of operation with specific conditions, i.e., using a mixture of inert gas and oxygen as breathing medium under high pressure. Although the air is a mixture of nitrogen and oxygen, the mixed gases, we generally refer, are gases artificially mixed with oxygen and inert gases in a specific ratio. The commonly used mixed gases include the mixtures of nitrogen and oxygen, helium and oxygen, hydrogen and oxygen, and other types of mixture gases.

After the inhaled oxygen dissolves into blood, it forms chemical compounds and is continuously consumed. When a body inhale a gas mixture that is different from the composition of air, the inert gas in the mixture will dissolve into blood and accumulate gradually, and reach a state called "saturated" in which the tension of the dissolved gas achieve an equilibrium with the partial pressure of the gas in the environment. At this time, if the air pressure of the environment falls (reduce pressure) or if the total air pressure of the environment does not fall but the concentration of the inert gas is reduced (replaced by other kinds of gases), then the tension of the previously mentioned dissolved inert gas is higher than the partial pressure of the gas in the environment, forming a state called "supersaturated." At this time, the dissolved gas in the tissue will diffuse into the environment and become free gas, which is known as "desaturation," until it reaches an equilibrium state again.

When the body breathes a new gas mixture of different components, it must go through three phases—a phase of start breathing, a stable phase, and a phase of stop breathing. To facilitate understanding, we refer to the case of breathing high-pressure air to explain how the inert gas enter into the body in the beginning, what is the distribution characteristics in the body, and how inert gases are released from the body during the phase of stop breathing or during the decompression process. These laws are put forward by Haldane in early twentieth century. Through summarizing the practical experience of air diving and carrying out animal experiments, they elaborated the movement rule of nitrogen, one of the inert gases, in the body when diving, and established a corresponding doctrine, forming a classical theory on the movement rule of the inert gases in vivo in the modern diving medicine. The subsequent development and improvements of the movement rule are based on this theory. This chapter focuses on the basic theory of Haldane doctrine and takes nitrogen as an example to clarify the movement rule of an inert gas during the pressurization and decompression processes.

Although breathing a small amount of hydrogen is somewhat different from breathing high-pressure hydrogen or nitrogen, but their basic rules are similar. The rules of breathing different inert gases are similar. Furthermore, if hydrogen enters into the body through gas routes such as intraperitoneal injection or drinking water containing hydrogen, then hydrogen will be rapidly absorbed by the body according to concentration gradient. Although the speed of absorption and the way of breathing in this case differ from normal breathing of hydrogen, there is still similarity in their way of releasing hydrogen.

2.2 Absorption of Hydrogen Gas

2.2.1 Saturation and the Degree of Saturation

In chemistry, the term "saturation" refers to a condition that the concentration of solute in the solvent reaches the maximum solubility limit.

In this book, the concept of "saturation" is different from its chemical concept. Moreover, it is not only used as a noun, adjective, but also used as a verb, which are listed below [1]:

1. If a gas is dissolved in the body fluid or tissue and its tension is equal to the partial pressure of the gas in vitro, i.e., the amount diffusing back out is equal to the amount diffusing in per unit time, then this state of dynamic equilibrium will be called "saturation."
2. If dissolved gases in a certain tissue reach saturation, then the tissue is called "saturated tissue."
3. When the partial pressure of a gas in the environment is higher than the tension of the gas in tissue, the in vitro gas will diffuse into the body by the pressure gradient. Over time, the tension of the gas in body will increase gradually until the pressure gradient disappears. This process is known as "saturation."

"Saturation" is a term used to indicate the saturation level in hyperbaric medicine. For the convenience of quantitatively expressing the different degrees of saturation and accurately calculate and compare them, a state of saturation is called complete saturation; a state that does not reach complete saturation is called a partial saturation state. If a partial saturation state reaches half of the expected full saturation, then it is called to be in a half-saturation state.

The saturation of inert gas in vivo is commonly expressed as a percentage. For example, 100 % saturation means complete saturation, 50 % means half-saturation, and so on. Correspondingly, the shortfall of saturation is also expressed as a percentage which means the unsaturated level. There is a reciprocal relationship between saturation and saturation shortfall. Their sum equals to 1.

2.2.2 Half-Saturation Time and Half-Saturation Time Unit

Half-saturation time is first proposed by Haldane. It refers to the time needed for "filling" half of the inert gas saturation shortfall that exists in certain types of tissue, and it is usually expressed as a symbol $t_{1/2}$. For example, "filling" half of the nitrogen saturation shortfall in blood or lymph needs 5 min, while "filling" half of the nitrogen saturation shortfall in the gray matter of the central nervous system needs 10 min; this 5 min is considered the half-saturation time of nitrogen in the blood, lymph, and some other tissues; and this 10 min is considered the half-saturation time of nitrogen in the gray matter of the central nervous system.

Half-saturation time is used as a timing unit to represent the saturation time for the inert gases, and it is called "half-saturation time unit." The half-saturation time unit (n) is equal to the quotient of the actual time (T) divided by the half-saturation time:

$$n = T / t_{1/2}.$$

For example, if the human body is exposed to the compressed air for 40 min, then for the blood and lymph ($t_{1/2}=5$), $n=40/5=8$ half-saturation time units; while for the gray matter of the central nervous system ($t_{1/2}=10$), $n=40/10=4$ half-saturation time units.

2.2.3 *Process of Saturation*

1. Completing saturation process through respiratory and circulatory system

After the body entering into a high-pressure environment, breathing gases of different compositions is similar to this case. If the partial pressure of nitrogen in the gas mixture is higher than the tension of dissolved nitrogen in the blood, then the gaseous nitrogen will spread rapidly through the alveoli into the blood and be taken by arterial blood to body tissues. The areas of the alveolar wall and the systemic blood capillary are very large and their walls are very thin; therefore, gas exchange between the lungs and body tissues can be said to be done in an instant. The main time of the saturation process is spent on the transport of gas through blood. It takes about 18 s for blood to circlulate inside the whole body. After the gas is delivered by blood to the tissues, the regurgitant venous blood will reengage with the alveoli. By this time, the pressure gradient of nitrogen between the mixed gases and the blood has been reduced compared to that in the previous circulation, and the diffusion of the inert gas from the alveoli into the blood and from the blood to the tissues is also decreased compared to that in the previous circulation. Repeatedly, over time, the tension of an inert gas within the tissues will reach equilibrium with the partial pressure of the gas in the environment, and the pressure gradient of the inert gas among alveolar air, blood, tissues, and venous blood will all disappear.

2. The growth rate of saturation reduces exponentially when the number of circulation or time units increases

Under atmospheric pressure, the tension of nitrogen in the body has been in a state of equilibrium with the partial pressure of nitrogen in the air. For a general adult male, the dissolved nitrogen in his body is approximately 1000 ml. Among this amount, the dissolved nitrogen in the blood is about 39 ml, which accounts to approximately 4% of the total dissolved nitrogen in the body. When the body has just entered the high-pressure, the saturation shortfall is 1. After a cycle of the systemic blood circulation (18 s), the body will increase the nitrogen saturation by 4%

Table 2.2 Nitrogen saturation calculated with blood circulations as timing unit

Order of time unit (n)	Cumulative time (s) ($n \times 18$)	Saturation completed ($4\% \times (1-4\%)^{n-1}$)	Saturation shortfall ($(1-4\%)^n$)	Cumulative saturation ($1-(1-4\%)^n$) (%)
1	$1 \times 18 = 18$	$4\% \times (1-4\%)^0 = 4\%$	$(1-4\%)^1$	4.000
2	$2 \times 18 = 36$	$4\% \times (1-4\%)^1 = 3.84\%$	$(1-4\%)^2$	7.840
3	$3 \times 18 = 54$	$4\% \times (1-4\%)^2 = 3.68\%$	$(1-4\%)^3$	11.526
4	$4 \times 18 = 72$	$4\% \times (1-4\%)^3 = 3.54\%$	$(1-4\%)^4$	15.064
5	$5 \times 18 = 90$	$4\% \times (1-4\%)^4 = 3.40\%$	$(1-4\%)^5$	18.461
6	$6 \times 18 = 108$	$4\% \times (1-4\%)^5 = 3.26\%$	$(1-4\%)^6$	21.722
7	$7 \times 18 = 126$	$4\% \times (1-4\%)^6 = 3.13\%$	$(1-4\%)^7$	24.853
8	$8 \times 18 = 144$	$4\% \times (1-4\%)^7 = 2.88\%$	$(1-4\%)^8$	27.858
9	$9 \times 18 = 162$	$4\% \times (1-4\%)^8 = 2.77\%$	$(1-4\%)^9$	30.744
10	$10 \times 18 = 180$	$4\% \times (1-4\%)^9 = 2.66\%$	$(1-4\%)^{10}$	33.514
11	$11 \times 18 = 198$	$4\% \times (1-4\%)^{10} = 2.55\%$	$(1-4\%)^{11}$	36.173
12	$12 \times 18 = 216$	$4\% \times (1-4\%)^{11} = 2.45\%$	$(1-4\%)^{12}$	38.726
13	$13 \times 18 = 234$	$4\% \times (1-4\%)1^2 = 2.35\%$	$(1-4\%)^{13}$	41.527
14	$14 \times 18 = 252$	$4\% \times (1-4\%)^{13} = 2.26\%$	$(1-4\%)^{14}$	43.527
15	$15 \times 18 = 270$	$4\% \times (1-4\%)^{14} = 2.17\%$	$(1-4\%)^{15}$	45.786
16	$16 \times 18 = 288$	$4\% \times (1-4\%)^{15} = 2.08\%$	$(1-4\%)^{16}$	47.955
17	$17 \times 18 = 306$	$4\% \times (1-4\%)^{16} = 2.08\%$	$(1-4\%)^{17}$	50.000
⋮	⋮	⋮	⋮	⋮
34	$34 \times 18 = 600$		$(1-4\%)^{34}$	75.000
⋮	⋮	⋮	⋮	⋮
51	$51 \times 18 = 900$		$(1-4\%)^{51}$	87.500
⋮	⋮	⋮	⋮	⋮
102	$102 \times 18 = 1800$		$(1-4\%)^{102}$	98.437

and the remaining saturation shortfall is "$1-4\%$;" the second cycle of circulation will complete 4% of the remaining shortfall, i.e., $4\%\ (1-4\%)$, and the remaining shortfall is $(1-4\%)^2$, and so on. Thus, we can see that the saturation that the latter circulation gained is diminishing exponentially compared to the previous one (Table 2.2). This is repeated until it is completely saturated.

If we use the half-saturation time as the timing unit, then the saturation of the inert gas gained in the first time unit is 50% and the saturation shortfall is 50%. In the second time unit, 50% of the saturation shortfall left in the first time unit is "filled," i.e., $50\% \times 50\% = 25\%$. The cumulative saturation that these two time units gained is 75%, and at this time, the saturation shortfall is 25%. In the third time unit, 50% of the saturation shortfall left in the second time unit is "filled," i.e., $25\% \times 50\% = 12.5\%$. The cumulative saturation reaches 87.5%, and by now the saturation shortfall is 12.5%, and so on. With the increase of the number of time units, the cumulative value of the saturation increases and the saturation shortfall left decreases (Fig. 2.1).

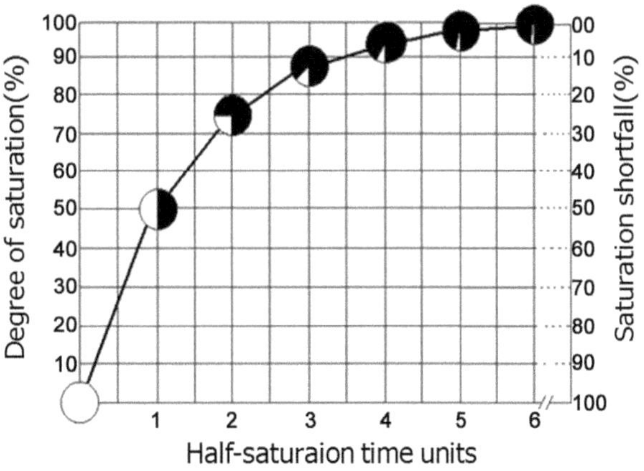

Fig. 2.1 The saturation growth diagram of inert gas (within the *circle* on the curve, the *blank* indicates the proportion of saturation shortfall, and the *blackened* indicates the proportion of the saturated)

In theory, to approach complete saturation needs numerous time units. But in general, when the saturation reaches 98% after six time units, we will deem this saturation as 100% saturation. The specific time of $t_{1/2}$ in different tissues is different. However, different tissues gained the same proportion of saturation within these six time units for the shortfall that inert gas left in the tissues.

The total saturation(s) cumulated after a certain number of time units (n) can be calculated using the following formula:

$$S = 1 - (1 - 50\%)^n.$$

Typically, this formula is simplified as $S = (1 - 0.5^n) \times 100\%$.

According to this formula, the cumulative value of nitrogen saturation after different time units of exposure can be calculated in detail (Table 2.3).

3. Inert gas in different tissues have different saturation velocity

If parameters of nitrogen dissolution in every tissue and in the blood are the same and the blood perfusion of every tissue is very abundant and smooth, then it will cost only 5 min to accumulate 50% of nitrogen saturation in the whole body (see Table 4.2). However, components of body's various tissues vary from the blood, the solubility coefficients of nitrogen in different tissues differ greatly, and the blood perfusion statuses among different tissues also have a great difference. Thus, the half-saturation time cannot be the same in different tissues; that is, saturation velocities are different in different tissues.

If a tissue is high in lipid, then because the solubility of inert gas in lipid is higher than in water and the tension of gas in tissue rises slowly, so its half-saturation time is long (slow); if a tissue has a rich blood supply, then because it can dissolve more inert gas per unit of time, the tension of gas in tissue rises more quickly, so its half-saturation time is short (fast). For tissues high in lipid and rich in blood supply, their half-saturation time may not be long; while for tissues low in lipid but less in blood supply, their half-saturation time may not be short. Of course, for tissues with more lipid and less blood supply, their half-saturation time will be very long ("slow tissue");

Table 2.3 Nitrogen saturation calculated with half-saturation time units as timing unit

Half-saturation time units (n)	Growth rate of saturation ($50\% \times (1-50\%)^{n-1}$)	Saturation shortfall $(1-50\%)^n$	Cumulative saturation[a] $(1-(1-50\%)^n)$
1	$50\% \times (1-50\%)^0 = 50\%$	$(1-50\%)^1 = 50\%$	$1-(1-50\%)^1 = 50\%$
2	$50\% \times (1-50\%)^1 = 25\%$	$(1-50\%)^2 = 25\%$	$1-(1-50\%)^2 = 75\%$
3	$50\% \times (1-50\%)^2 = 12.5\%$	$(1-50\%)^3 = 12.5\%$	$1-(1-50\%)^3 = 87.5\%$
4	$50\% \times (1-50\%)^3 = 6.25\%$	$(1-50\%)^4 = 6.25\%$	$1-(1-50\%)^4 = 93.75\%$
5	$50\% \times (1-50\%)^4 = 3.12\%$	$(1-50\%)^5 = 3.12\%$	$1-(1-50\%)^5 = 96.8\%$
6	$50\% \times (1-50\%)^5 = 1.56\%$	$(1-50\%)^6 = 1.56\%$	$1-(1-50\%)^6 = 98.437\%$
...	...	...	...

[a] Commonly used formula: $S = (1-0.5^n) \times 100\%$

for tissues with less lipid and more blood supply, their half-saturation time will be very short ("fast tissue").

2.2.4 Theoretical Tissue Compartments

The body tissues were classified into five categories by Haldane depending on the nitrogen half-saturation time in the tissues, and such tissues are termed as the theoretical tissue compartment.

According to the length of half-saturation time, Haldane classified the body tissues into the following five categories (I–V) of theoretical tissue compartments:

Type I tissue compartments: $t_{1/2} = 5$ min, and also known as 5 -min tissues, including blood, lymph, etc.

Type II tissue compartments: $t_{1/2} = 10$ min, and also known as 10 -min tissues, including the gland, the gray matter of the central nervous system, etc.

Type III tissue compartments: $t_{1/2} = 20$ min, and also known as 20 -min tissues, including muscle, etc.

Type IV tissue compartments: $t_{1/2} = 40$ min, and also known as 40 -min tissues, including lipid, the white matter of the nervous system, etc.

Type V tissue compartments: $t_{1/2} = 75$ min, and also known as 75 -min tissues, including tendons, ligaments, etc.

The five theoretical tissue compartments have different half-saturation time, but as long as they were giving the same number of time units, they will achieve the same saturation. After six units of the respective half-saturation time unit, the saturation of all the five theoretical tissue compartments can reach 98%. We can also calculate the time required for each theoretical tissue compartment to achieve "complete saturation" according to their half-saturation time (t_s). The formula is $t_s = t_{1/2} \times 6$.

According to this calculation formula, the time required for the five theoretical tissue compartments to achieve "complete saturation" were, in order, $5 \times 6 = 30$ min, $10 \times 6 = 60$ min, $20 \times 6 = 120$ min, $40 \times 6 = 240$ min, and $75 \times 6 = 450$ min. Thus, it is visible that the longer the half-saturation time of a tissue is, the longer time is needed to achieve "complete saturation." But if the body is exposed to air for 450 min, then nitrogen in all kinds of theoretical tissue compartments will achieve "complete saturation."

From the above discussion, we can grasp two important messages. First, when human body breathes a concentration of hydrogen (e.g., 2%), the hydrogen concentration in the human body increases gradually; second, different tissues differ in the increase rate of hydrogen concentration. According to the law of theoretical tissue compartment, the concentration in the blood increases first and then brain tissue while the increasing speed in some slow tissues is very slow. The maximum saturation concentration in the blood can be reached in 30 min, and continue to breathe hydrogen will not increase the hydrogen concentration in the blood. However, the brain requires 60 min to achieve the same maximum saturation concentration and other tissues require more time.

2.3 Release of Hydrogen Gas

With regard to the whole body, when a body exposed to the high pressure returns back to atmospheric pressure and within a period of time, the tension of inert gas that have been dissolved in the body in the high pressure is higher than the external partial pressure of the gas. As a result, the inert gas will diffuse into the outside according to the pressure gradient until the tension of inert gas reaches an equilibrium with the partial pressure of the gas. This process is termed as desaturation of inert gas. The hydrogen partial pressure in vivo will exceed that in the external respiratory gas after the end of hydrogen breath or other means of hydrogen supply, the in vivo hydrogen will be released to the outside in a manner similar to circulation and respiration. Because hydrogen has a very large diffusion capacity, the proportion of hydrogen released through the skin cannot be ignored.

Haldane believed that the inert gas desaturation and saturation only differ in the opposite direction of diffusion; that is, the direction of dissolved gases carried by the blood when desaturation is not from the lungs to the tissues, but from the tissues to the lungs. But, in fact, in addition to the opposite direction, the time needed for desaturation is much longer than that for saturation. There are two main reasons: (1) When desaturation happens, gas will diffuse from the liquid phase to the vapor phase and thus the gas molecules will be bound by the liquid. If the liquid contains substances like colloidal protein, then the binding effect will be more obvious. (2) In order to ensure safety, the desaturation process also needs to be controlled by the safety factor of supersaturation. As to the law that inert gas takes advantage of the pressure gradient to diffuse, it is the same to that of saturation, that is:

1. The completion desaturation process is also through the functional activities of the respiratory and circulatory systems. Other ways of desaturation can also

be ignored. Many factors affect desaturation rate through directly or indirectly affecting the activities of the respiratory and circulatory systems.

2. Tissues that are faster to be saturated are also faster to be desaturated; tissues that are slower to be saturated are also slower to be desaturated. Customarily, the former is called fast tissues and the latter the slow tissues.
3. Completion of 50 % desaturation needs a half-saturation time unit; completion of 98 % desaturation ("complete desaturation") needs six time units. In short, the longer the residence time at a lower pressure, the more complete the desaturation is (Fig. 4.2).
4. The degree of desaturation is also according to the order of time units. The desaturation degree of the latter time unit diminishes exponentially compared to the former.

The saturation and desaturation law of inert gas can be used to interpret the change rules and characteristics of hydrogen in vivo. For example, due to a slower blood circulation in human than in small animals, the increase in rate of hydrogen concentration in human blood is lower than that in small animals at the same dose of hydrogen. If hydrogen breathing or hydrogen injection of same amount is stopped when its concentration in the blood and tissues reaches the maximum, the decline rate of its concentration in human body is also relatively slower than that in smaller animals. Both the increasing and reducing processes of the hydrogen concentration are in line with the exponential change law of the inert gas saturation and desaturation.

Reference

1. Tao HY. Diving medicine. Shanghai: Shanghai Science and Technology Press; 2010. pp. 79–90.

Chapter 3
Biological Safety of Hydrogen

Qiang Sun, Wenjie Han and Atsunori Nakao

Abstract Based on three main reasons, the biosafety of hydrogen is very high: The first is evidence from medical research on hydrogen diving, the second is that hydrogen is an endogenous gas, and the third is direct research on the biosafety of hydrogen. Research on hydrogen diving, combined with human trials on hydrogen diving, proved that hydrogen is very safe for humans to breathe. Since a certain level of hydrogen is produced by *Escherichia coli* in the large intestine of normal humans, hydrogen can be considered an endogenous gas. So far, no clinical evidence has been found that hydrogen can be harmful for the human body. Published data from the EU and the US government on the biosafety of hydrogen showed that hydrogen has no acute or chronic toxicity on the human body under normal pressure. Despite this, any substance that can produce biological effects on the human body has the potential of destroying homeostasis, which may be harmful. Although the biosafety of hydrogen is very high, we still cannot assure that hydrogen has no side effects on the human body.

Keywords Hydrogen · Biosafety · Endogenous gas · Diving · Narcotic effect

3.1 Studies in Diving Medicine

Hydrogen was first named by the famous French chemist Lavoisier, who was also the forerunner of the study of the physiology of hydrogen. Since hydrogen and oxygen can be chemically combined into water, Lavoisier named hydrogen using *Hydro* as the root which means elements of water. In Japanese and Korean

A. Nakao (✉)
Department of Emergency, Disaster and Critical Care Medicine, Hyogo College of Medicine, Nishinomiya, Japan
e-mail: atsunorinakao@aol.com

Q. Sun
Department of Hyperbaric Medicine, Navy General Hospital, Fucheng Road, Beijng, China

W. Han
Department of Cadre Ward, Navy General Hospital, Fucheng Road, Beijng, China

X. Sun et al. (eds.), *Hydrogen Molecular Biology and Medicine*,
DOI 10.1007/978-94-017-9691-0_3

languages, hydrogen is translated directly into elements of water. So in the market, elements of water equal hydrogen water. Our chemistry named hydrogen as "Qing," since it has a pronunciation similar to light in Chinese and it is a kind of gas.

As early as 1789, the famous chemist Lavoisier and Sequin performed animal studies using hydrogen as breathing medium. In the experiment, Lavoisier put guinea pigs into a bell-shaped glass container; hydrogen was added into the container while nitrogen and oxygen were maintained above a certain level to keep them alive. No adverse effect was found after 8–10 h. [1].

In 1937, English scientists Case and Haldane carried out a human study in which hydrogen was added to breathing gas during diving. The subjects were exposed under a pressure of 1.1 MPa when breathing hydrogen–oxygen mixed gas for 5 min, and no adverse effect was found [2]. In 1941, the former Soviet Union scientist Lazarev conducted an experiment on mice under a pressure of 9.1 MPa; mice breathed a gas mixture of hydrogen, nitrogen, and oxygen for 3 min, and then survived after 1-h decompression [3]. These early studies could draw a preliminary conclusion that it is safe for humans and animals to breathe a gas mixture of hydrogen and oxygen.

In the early stage of diving experiments on breathing oxygen and helium conducted by US scientists, a hydrogen mixture was attempted in diving studies in many countries since it was very difficult to obtain helium at that time. The most famous example was the experiment conducted by Arne Zetterstrom (Fig. 3.1), a diving technical engineer in the Sweden navy, who tried a gas mixture of hydrogen and oxygen during diving [4].

Fig. 3.1 Arne Zetterstrom (1917–1945), diving technical engineer in Sweden

Hydrogen–oxygen mixed gas is explosive, but no explosion will occur when oxygen concentration is lower than 4 %. Under normal pressure, life will not survive when oxygen concentration is 4 %. However, the partial pressure of oxygen can be elevated to 16 KPa. In 1944, Zetterstrom invented a way of producing hydrogen and oxygen mixture from air; the concentration of oxygen in this mixture is 4 % and explosion can be avoided. The hydrogen and oxygen mixture was then applied in diving. Zetterstrom first replaced used air by a nitrogen and oxygen (4 %) mixture as breathing gas at a depth of 30 m, then nitrogen was replaced by the same percentage of hydrogen, and he successfully dived to a depth of 100 m. He became extraordinarily excited at this depth (anesthetic effect of high-pressure hydrogen), and his voice also changed obviously, which influenced the communication with people on the surface.

On August 7, 1945, to prove that hydroxide diving techniques play an important role in submarine rescue, Zetterstrom successfully dived to a depth of 161 m in the biggest submarine rescue ship *Belos* by breathing hydroxide gas. Unfortunately, since there was a lack of a standard decompression procedure and faulty operation by surface staff, Zetterstrom directly ascended to the surface without decompression in the 50-m decompression stop. Zetterstrom died due to severe anoxia and decompression sickness [5].

Since 1960, diving medicine, especially the improvement of saturated diving technology, promoted animal experiments on hydroxide diving to be conducted again by US, English, French, Russian, and Swedish scientists [6]. Between 1960 and 1970, the improvement of hydrogen-producing technology had a greater price advantage; replacing hydrogen with helium aroused high attention again. During this period of research on hydroxide diving, animal experimentation had reached a depth of 1000 m and breathing hydrogen exposure time up to 24 h. Human experimentation has reached a depth of 60 m and exposure time 10–20 min. In the 1970s, Edel conducted simulation diving experiments on hydroxide. He not only studied the effects of hydrogen on the body but also included hydroxide diving safety procedures, decompression plans, and respiratory gas conversion technology [7]. The greatest achievement at that time was the finding that hydrogen could prevent high-pressure nervous syndrome (HPNS), which is one of the most important advantages of breathing hydrogen. Breathing helium at a depth over 300 m could cause HPNS, which is one of the biggest obstacles during increasing diving depth [8].

With continuous ocean development, greater depths of diving operations are needed, which promote the study of hydroxide diving. Throughout the world, France has been in the leading position with respect to hydroxide diving study. COMEX is a French company that once led research on diving medicine and technology. In the early 1980s, the company began to carry out a dive plan using a deep-sea gas mixture Hydra (Hydra means hydrogen, as hydrogen was the main gas in the mixture). The dive plan included animals, human simulation, and field experiments, consisting of three sections which were security check, medical physiology, and developing diving equipment. Hydra III–VII simulation experiments were conducted in which a routine dive at a depth of 75–90 switches to hydroxide saturation diving at a depth of 520 m. During Hydra V, hydrogen was first used in saturation diving at a depth of 450 m; the divers completed mechanical connection, underwater cutting, and several diving operation tasks [9, 10].

In June 1983, COMEX started the Hydra plan. Under the leadership of Deulaze, president and concurrently diver of the company, Hydra III diving experiments were conducted at a depth of 91 m near Marseille; the breathing hydrogen gas mixture consisting of 95 % oxygen was 5 %. No anesthetic effects of hydrogen were found, heat dissipation effects were similar to helium, and breathing resistance was lower than helium. In November 1983, Hydra IV human simulation experiments were conducted at a depth of 300 m, six subjects breathing a gas mixture of hydrogen, oxygen, and helium, the gas ratio of which was 74:24:2, respectively. Psychology examination found that the visual reaction time, calculating ability, and memory effects were less than those at a depth of 80-m air diving as well as those at a depth of 240-m hydroxide diving (gas ratio of hydrogen to oxygen less than 98:2). At a depth of 180 m, breathing a gas mixture of hydrogen to oxygen (98:2), the anesthetic effect began to appear; hydrogen anesthesia was different from nitrogen narcosis, illusion being the main symptom, the anesthesia level of which was also lower than nitrogen narcosis. Changes in heart rate during hydroxide diving are less than in helium–oxygen diving. After hydrogen simulation diving, indicators of blood, urine, nervous system, and respiratory function were normal. Simultaneously, hydroxide diving animal experiments were conducted; 40 mice were exposed at a depth of 600 m for 40 h, after decompression to ambient pressure, and no abnormalities were found on heart, liver, and lung histology examination [11].

In May 1985, Deulaze carried out the Hydra V experiment; the subjects were divided into two groups, with exposure to a depth of 450 m. The physiological role of hydrogen was studied, underwater capabilities of divers were tested, and the gas mixture ratio was systematically studied at the same time. It was found that the optimal ratio of hydrogen to helium and oxygen was 54:45:1, and during the operation breathing was smooth and no pressurized joint pain appeared [12].

These series of experiments further confirmed that hydrogen is not harmful to the body, while establishing the best safety standards for hydrogen gas mixture preparation.

In November 1986, Fructus et al. carried out a Hydra VI simulated dive, in which eight divers participated in simulated diving at a depth of 520 m 25 times while the water temperature was 4 °C [10]. In February 1988, COMEX conducted Hydra VIII in the Mediterranean applying hydrogen mixed gas in saturation diving experiments. Six divers took part in the experiment. In February 21, compression started until a depth of 520 m was reached; in the following 6 days, six divers conducted several diving operations [13]. On September 1989, the COMEX company conducted Hydra IX in which divers conducted saturation diving simulation for 14 consecutive days. The purpose of the experiments was to investigate the minimum and maximum depth limit in using hydrogen, the long-term (49 days) effect of high-pressure hydrogen environment exposure on human physiological function, long-time isolated confinement effects on the behavior of the human spirit, and so on. This item completed research plans including diving medicine, neurophysiology, psychology, ventilation and cardiovascular function, biochemical, thermal, and fluid balance, diving decompression procedures, expert systems, etc. [14].

In the 1990s, COMEX successively carried out experiments of saturation diving under 680 m with excursion diving under 701 m (Hydra X) and simulated human saturation diving under 750 m (Hydra XI) [15]. The maximum depth recorded on human exposure to high pressure was 750 m. Currently, COMEX Hydra XII experiments are being carried out; divers dive to a depth of 210 m for 28 dives, including four times for helium–oxygen diving and four for hydrogen helium–oxygen diving. Throughout the experiment, divers breathed heliox in the deck saturation chamber, while breathing a hydroxide mixture during underwater operations. These experiments explored the feasibility of existing equipment for hydroxide diving and closed respiratory system for underwater breathing gas conversion (helium gas mixture converted to hydrogen gas mixture) situation. Each dive lasted for 2–6 h; both analytical thinking and operating capacity of the divers were normal when they breathed the hydrogen gas mixture [16].

In the Hydra plans, Gardette successfully conducted experiments on 110 mice using a hydrogen and helium–oxygen mixture, exposed to a condition of 2000 m for 12 continuous days [10]. It was found that when mice were pressurized with hydroxide at 1800 m, HPNS appeared obviously, but no such syndrome appeared when they used a hydrogen and helium–oxygen mixture instead. These results indicated that hydrogen could alleviate HPNS.

In recent years, more countries have begun to pay close attention to hydroxide diving gradually. The US Navy studied the physiological effects of hydrogen on the human body under high pressure; it was confirmed that hydrogen is harmless, and that it could not only effectively reduce breathing resistance under high pressure but also alleviate the HPNS. US scientists also carried out a special experiment related to hydroxide diving: the biology decompression method for hydroxide diving. Methanogenic bacteria could be capable of converting carbon dioxide and hydrogen into methane and water, through which the partial pressure of hydrogen in vivo could be reduced promptly. They used the above principle to accelerate the decompression of hydroxide diving. In 1998, Kayar reported that decompression time can be effectively shortened through injecting methanogenic bacteria into animal gut; meanwhile, the incidence of decompression sickness was also reduced [17]. More importantly, the following study found that the methanogenic bacteria belong to the normal intestinal flora, and decompression time can be shortened to a certain extent by itself in normal circumstances [18]. If we implement this technology in future diving, decompression time will be further shortened, which is an important direction for future diving medical research.

Since there exist many technical obstacles, hydrogen diving is currently at the research stage. However, the important progress we have made, particularly the knowledge of the physiology of hydrogen diving, remains very valuable.

3.1.1 High-Pressure Hydrogen Gas Has No Toxic Effects

The results of the present study clearly show that hydrogen has no toxic effects on the body at any pressure. Since Lavoisier began to study the role of hydrogen on the body from 1789, almost all research involves the safety issue of hydroxide diving. Although it was once thought that hyperbaric hydrogen may have toxic effects on the body, it turned out to be wrong in later research. Especially when the series of human hydroxide diving experiments were carried out by Sweden, France, etc. successfully, such as the experiments conducted by France from 1988 to 1989, human hydroxide diving time added up to 7200 h. These experiments further explained that it is safe for humans to breathe hydroxide when diving. Since high-pressure hydrogen has no toxic effects, a small dose of hydrogen gas would be safer.

3.1.2 High-Pressure Hydrogen Gas Has a Certain Narcotic Effect

Physiologically inert gases tend to have a certain narcotic effect at high pressure, most notable of which is the narcotic effect of nitrogen. As long as the ambient pressure increases by four times, the narcotic effect will occur in many people. When the pressure increases up to ten times, a very serious narcotic effect will occur in most people, which is also an important reason why air diving should not exceed 60 m. Animal experiments suggest that high-pressure hydrogen also has a certain narcotic effect, which can resist HPNS to a certain extent. This happens to be a problem helium diving could not solve, which becomes one of the advantages of hydrogen toward deep sea diving. The narcotic effect of hydrogen ranks between nitrogen and helium, the relative narcotic effect of nitrogen, hydrogen, and helium being 4.3:2.3:1. The narcotic effect of hydrogen is about 54 % of nitrogen, but higher than helium and neon. People took advantage of this characteristic of hydrogen to reverse the excessive excitatory effects on the body caused by high pressure. Recently, animal and human studies showed that HPNS could be antagonized by hydrogen. HPNS will not occur in humans when they breathe 4.6-MPa (46-atm) hydrogen, nor is working ability affected [19, 20].

The main signs of the narcotic effect of high-pressure hydrogen are hallucinogenic effects, which have a high impact on the cognitive ability of divers; even when the body is exposed to 1.9-MPa high-pressure hydrogen environment, this effect will occur. In 1974, Edel performed the first study on the narcotic effect of hydrogen, in which four subjects dived once, under the conditions of 0.7-MPa breathing nitroxide, heliox, and hydroxide for 120 min. Results of intelligence and operational skills showed that nitroxide has the most obvious influence on operational skills, while no significant effect of heliox and hydroxide was found. In the experiments of Hydra IV, Hydra A, and Hydra V, conducted in 1984, Carioz once investigated the narcotic effect of hydrogen on human bodies; the results showed no obvious changes in operation skills and visual reaction time when breathing 2.45-MPa hydrogen. Human studies also showed that there exist obvious different narcotic sensitivities among individuals [21, 22].

3.1.3 The Diffusion Rate of Hydrogen Is Very Fast

Since the diffusion rate of hydrogen is faster than air, it is easier for hydrogen to pass through narrow pores which is beneficial for the pressure adjustment between inner and outer side of gas chamber such as eardrum room. Hydrogen's high diffusion ability provides a guarantee for exerting its biological effects; hydrogen can enter into any sites of the body, and any parts of the cell, such as the nucleus and mitochondria. Hydrogen possesses this advantage over many other drugs.

3.1.4 Body Conducts Heat Faster Under Hydrogen Condition

Both thermal conductivity and heat capacity of hydrogen are larger than nitrogen and helium. Therefore, when breathing hydroxide gas, more heat is lost easily from the lungs, which is a major limiting factor for hydrogen diving. Experiments show that at 310-m depth in a hydroxide environment, body heat loss is very fast; even if engaged in heavy physical operation, metabolic heat production cannot compensate the loss of body heat. So hydroxide diving requires external heating. In a hydroxide environment, it is appropriate to maintain cabin temperature within 34 °C. In addition, the breathing gas should be heated to a certain extent to make divers feel comfortable. However, there exists a debate over the optimum temperature. Smith et al. found that, under the condition of 1.1–10.1 MPa, the speed of respiratory cooling and convection cooling of breathing hydroxide is 38 % and 32 % higher than that of helium oxide, respectively. They proposed that a comfortable temperature range of hydroxide diving should be 31.1–31.6 °C [23].

3.1.5 Effects of Breathing Hydrogen on Voice

Current researches focus on the voice effect of helium; only a few researches have reported on the voice effect of hydrogen. In the high-pressure hydrogen environment, the human voice changes significantly, such as higher tone, with nasal voice, lower clarity, and being difficult to understand, and so on. During the early hydrogen diving experiments conducted by Swedish engineer Arne Zetterstrom, voice change problem was found which significantly affected the language communication with the surface. His later death accident was caused indirectly by the voice change.

3.1.6 Effects of High-Pressure Hydrogen on Respiratory System

Hydrogen gas is the least dense gas in the universe. When breathing hydroxide, the respiratory resistance is lower; therefore, the work of breathing and physical exertion reduced which helped improve diver's operational efficiency. Dougherty found that the maximum voluntary ventilation of breathing pure oxygen was equivalent to that of 40 % helium–oxygen; other respiratory indicators changed significantly, mainly due to higher relative density of oxygen, suggesting that ventilation should be larger when breathing hydrogen mixed gas [24, 25]. However, from 4.6-MPa helium oxygen diving simulation experiments conducted by Giry, no significant improvement of ventilation function was found; on the contrary, ventilation function was reduced which might be caused by the narcotic effect of high-pressure hydrogen on the respiratory center [26]. Dahlback found that under 1.3-MPa conditions, functional lung impedance, when breathing 98 % hydrogen and 2 % oxygen, was decreased by 35 % compared to that of 98 % helium and 2 % oxygen, which could reduce the load of the respiratory muscles and then improve the work capacity of deep divers. In addition, breathing hydroxide under the conditions of 1.3 MPa, lung capacity slightly increased, but Giry's experimental results did not find this [27].

3.1.7 *Effects of High-Pressure Hydrogen on Circulation System*

Gennser et al. studied the effect of hydrostatic pressure, hydrogen, nitrogen, and helium on rat atrial rate at the cellular level. The results showed that 15-MPa hydrostatic pressure reduced the isolated rat atrial spontaneous frequency by 30.6 ± 7.2 %; if perfused with high-pressure hydrogen (hydrogen partial pressure: 4.9, 9, and 14 MPa) saturated solution, the rat atrial spontaneous frequency increases with the increase in hydrogen partial pressure [28]. Some researchers also found nitrogen to be twice as strong as hydrogen over the effect of alleviating bradycardia, but hydrogen is five times stronger that helium. The same effect was found in 5-MPa nitrogen and 9-MPa hydrogen toward reversing bradycardia. Hydra V experiments showed that hydrogen could resist the effect of hydrostatic pressure. Giry believes that the cardiovascular changes found when breathing high-pressure hydrogen gas mixture might be caused by inhibition of the excitability of the parasympathetic nervous system by hydrogen [29].

Diving medicine investigated the biological effects of hydrogen through exposing animals and humans to very high pressures. Since there exists a relationship between the effect of gas and the concentration of dissolved gas, and the concentration of dissolved gas is related to its partial pressure, the biological effects of high-pressure hydrogen focus on the effects of hydrogen under high pressure, which is different from our current concept of the biological effects of hydrogen. However, by reviewing the relevant knowledge from diving medicine, we can clearly know that even at very high-pressure conditions, hydrogen gas is still very safe for the human body. So when the dosage of breathing hydrogen is far below the diving condition thousands of times, the biosecurity of hydrogen can be assured.

3.2 Hydrogen Is an Endogenous Gas

Bacteria of human large intestine can produce hydrogen, which at least proves that hydrogen is an endogenous gas existing in the normal environment of human cells. Although the evidence is not enough to prove the safety of hydrogen, it provides some clues.

3.2.1 Studies of Intestinal Gas

Studies on gas production and composition of the large intestine have a long history; the composition and sources of gas were the main focus in early times. Even though we know little about the specific details of the gas sources of the large intestine, data about bacteria of the large intestine, which produce hydrogen, are adequate.

According to the literature, the first attempt was to measure gas of the large intestine in 1868. Ruge E, a German scholar, founder of the large intestine gas discipline, utilized a special chair to confine subjects; the anus was connected to a glass tube and gas was collected through the draining water by the gathering of gas law [30]. In 1942, Beazell and Ivey measured gas collected from 24 healthy people in 24 h; they found that 1 day's gas production of one person was about 380–655 ml. Later, Kirk proved cellulose in foods can increase gas production [31]. Steggerda collected gases with an average of 360 ml every day from some food (boiled eggs, beef, and fine apple juice), which was originally thought not capable of producing gas; further analysis found that it was composed of 7.4 % methane and 19.8 % hydrogen, which initially proved that methane and hydrogen are from coliform bacteria. In the following 7-day test, researchers observed the effects of different types of food in the body gas produced through adding different types of legumes into food, without changing the total energy intake from overall protein, fat, and carbohydrates [32]. The results found that small-molecule carbohydrates such as monosaccharides, disaccharides, and oligosaccharide can contribute to gas production. For example, if choosing pork and soy as the main food, gas production can be increased to 4.2 times. Using a constant perfusion technique, Levitt et al. proved that hydrogen can be produced by all healthy human large intestines, most of the hydrogen generated being entirely dependent on the bacterial fermentation of food components [33].

Gas produced by human colon is mainly composed of hydrogen, carbon dioxide, methane, nitrogen, oxygen, and other trace gases. Nitrogen and oxygen are mainly taken in from the mouth through swallowing into the digestive tract. Gas concentrations of hydrogen in the large intestine can be up to 74 %. Carbon dioxide can be produced by coliform bacteria, which digest dietary ingredients escaping from the human digestive enzymes or endogenous "food" dropping off the colonic mucosa. Some trace gases are really rare. For instance, one research found that the concentration of hydrogen sulfide in the human large intestine was only 1.06 μM, the concentration of methyl mercaptan only 0.21 μM, and the concentration of dimethyl sulfide only 0.08 μM. These trace gases are not only highly diffusible but also easily degradable, resulting in the final concentration of these trace gases released through the anus perhaps being lower than that in the large intestine.

The gas composition is not only affected by the bacterial species in the large intestine but also closely related to the physiological state of the host. As much as 30–40 % of carbon dioxide, hydrogen, and methane can be absorbed by colon mucosa into the circulatory system, and then released by the skin and respiration outside the body. Remaining gases such as hydrogen can be used for other bacteria, or eventually released through the anus. Therefore, the concentration of hydrogen may reflect the ratio of hydrogen producing and hydrogen-consuming bacteria [34].

Less than 20 % of carbohydrates in a typical western diet cannot be directly absorbed by the body; theoretically, 340 ml of hydrogen can be produced by bacteria from 1 g of glucose. Therefore, we can reckon that the maximum amount of hydrogen produced by bacteria in the human body is 13 l. But results from Strocchi and Levitt's test, perfusion with glucose, showed that each gram of glucose can only produce 80 ml of hydrogen instead of 340 ml [35]. So energy metabolism in bacteria is not so simple; there exist metabolic pathways that do not produce hydrogen while consuming energy substances. Hammer found that ingestion of 12.5 g lactulose will generate 50–200 ml of hydrogen per 6 h [36]. In fact, there is a huge difference in the amount of hydrogen produced according to different diet compositions and individuals. This difference reflects the very complex relationship and influence of factors existing between the human body and bacteria, for example, food type, composition and the amount of food, carbohydrate utilization ability of microorganisms in the large intestine, different number and location distribution of hydrogen producing or hydrogen consuming bacteria, efficiency of intestinal motility, pH, sulfide, and other environmental factors.

The National Aeronautics and Space Administration (NASA) once conducted a comprehensive study during the Apollo program on gas composition of the large intestine [37]. Most people might doubt the motivation of studying gas composition of the large intestine by NASA. In fact, this is the major obstacle related to security management inside the spaceship, since a high proportion of combustible gases are produced by astronauts themselves. If the gases are not properly handled, this could cause an explosion or fire in the spacecraft when these gases happen to ignite under certain circumstances. In addition, some ingredients in these gases are toxic, which may be harmful to the health of astronauts if no appropriate action has been taken. After careful analysis, NASA researchers found that gas compositions of the human large intestine are up to 400 species. The main gas components are nitrogen, hydrogen, carbon dioxide, methane, oxygen, and other odorless gases, plus some trace gases including ammonia, hydrogen sulfide, indole, skatole, volatile amines, volatile fatty acids, and other gases. Putting together all the malodorous gases, the volume can be no more than 1 % of the total volume. In the gas components, except nitrogen and oxygen, the vast majority of the gases are produced by intestinal bacteria.

3.2.2 The Biological Effects of Hydrogen in the Large Intestine

In the past, people knew little about hydrogen in the large intestine; most people thought that hydrogen was metabolic wastes from the large intestine bacterial metabolism and had no special effects. However, with the discovery of the biological effects of the hydrogen molecule, people gradually consider inducing colonic bacteria to generate hydrogen gas, which could exert a therapeutic effect on certain diseases.

Ohta's research group from the Nippon Medical School first proposed the issue; they tested breathing gas composition in healthy people after orally taking acarbose, and found that the concentration of hydrogen in the breathing gas was significantly increased but there was no effect on methane concentration. Acarbose is a class of drugs for treating diabetes whose mechanism is inhibition of intestinal epithelial absorption of glucose. Since the intestinal absorption of glucose is reduced, glucose will be transported to the large intestine through intestinal motility; bacteria of the large intestine can utilize glucose and produce a large amount of hydrogen. Thus, the common side effects of these drugs are abdominal distention and and excessive Farting. Early in the acarbose study, significant cardioprotective effects had been found, the molecular mechanism of which was not very clear. Since taking acarbose orally can promote the generation of hydrogen, and hydrogen has a protective effect against heart disease, they speculated that the reason why acarbose has cardioprotective effects was the generation of hydrogen [38]. Although this is very likely, this hypothesis still needs to be proved.

Since protective effects of endogenous hydrogen gas have been confirmed, then whether additional protective effect will appear after the producing of hydrogen is strengthened? Recently, researchers have found that orally administered cellulose or starch has a protective effect on the liver ischemia through inducing large intestinal bacteria to produce hydrogen [39]. Researchers provided rat cellulose and starch which cannot be digested directly for seven continuous days, and then detected the transaminase activity, which represents liver function or damage, the ratio of total glutathione and oxidized glutathione, which represents the degree of oxidative stress in liver. They found that the oral administration of cellulose or starch could promote the generation of hydrogen, improve liver function, and reduce oxidative stress after liver ischemia. The results proved that disease could be treated by promoting the generation of hydrogen gas. Dietary fiber was proved to be able to treat diseases many years ago, but the mechanism underlying this is yet to be found. There are a lot of resistant starch products in the market, although there are many explanations, but none is recognized. Promoting the large intestine to produce hydrogen is one mechanism that can clearly explain the effect of dietary fiber and resistant starch.

It remains controversial whether hydrogen of the large intestine has a biological effect. The body produces a huge amount of hydrogen, e.g., normal human bacteria have been reported to be capable of producing hydrogen 140 ml or even more which equals 10 l of hydrogen-saturated water. In many animal and clinical studies, hydrogen was given through injection or drinking at a dosage of 10–20 % of the above dosage; why is its effect not covered by that of endogenous hydrogen? There are two questions to be explained in this phenomenon. First, does endogenous hydrogen have an effect? Second, in which way does the relatively low dose of exogenous hydrogen exert its effect? This is a contradiction, which remains to be solved in the field of hydrogen research.

A recent study provides a reasonable explanation. The study observed the effect of oral lactulose-induced hydrogen produced by colonic bacteria in the animal model of Parkinson's disease (PD), healthy people, and PD patients [40]. Lactulose is one kind of disaccharide that cannot directly be absorbed by the body, but can be used to produce hydrogen gas in the large intestine by bacteria. Lactulose is widely used in clinical practice to treat constipation and hepatic encephalopathy.

End-alveolar breath hydrogen concentrations were measured in 28 healthy subjects and 37 PD patients, as well as in 9 rats after taking hydrogen water or lactulose. Six-hydroxydopamine (6-OHDA)-induced hemi-PD model was stereotactically generated in rats. The authors analyzed the effects of continuous and intermittent administration of 2 % hydrogen gas.

For intermittent administration of 2 % hydrogen gas, rats were supplied serially with 2 % hydrogen gas for 15 min and then room air for 45 min using a time controller. The 1-h cycle was repeated 12 times from 6 pm to 6 am to recapitulate the habit of drinking water once every hour in the dark. Each hydrogen administration protocol was started 1 week before the surgery. The control group was supplied with 2 % hydrogen gas continuously.

The results showed that lactulose increased breath hydrogen levels monophasically in nine rats.

However, oral lactulose and continuous administration of 2 % hydrogen gas had no obvious effect on ameliorating 6-OHDA-induced PD, whereas intermittent inhalation had effects on prevention of PD in rats. This study suggested that intermittent inhalation of hydrogen had a better effect than that of continuous inhalation. Marking effects of ad libitum administration of hydrogen water in a rat model of PD were found. Although lactulose also increases hydrogen levels in rats, lactulose and continuous inhalation of 2 % hydrogen gas have marginal effects on the prevention of 6-OHDA-induced PD.

Notably, since the release of hydrogen from the body is very rapid, drinking water is similar to intermittent inhalation of hydrogen. This study may explain why the human body can produce a large amount of hydrogen itself, but drinking hydrogen water or short inhalation of hydrogen may have a better effect, because the study found that intermittent administration of hydrogen had a better effect than that of continuous administration. Hydrogen is continuously produced in the body and its concentration is relatively stable, but the effect of drinking hydrogen water, intermittent inhalation, or injection of hydrogen are caused by rapid increasing of hydrogen concentration. Although the authors did not provide evidence at the molecular level, this is a relatively reasonable explanation.

Evidence of the treatment effect of endogenous hydrogen is from research by Harvard University, Boston Children's Hospital. This study examined whether H(2) released from intestinally colonized bacteria could affect Concanavalin A (ConA)-induced mouse hepatitis [41]. Systemic antibiotics significantly decreased the level of H(2) in both liver and intestines along with the suppression of intestinal bacteria. As determined by the levels of aspartate aminotransferase (AST), alanine aminotransferase (ALT), tumor necrosis factor (TNF)-alpha, and interferon (IFN)-gamma in serum, suppression of intestinal bacterial flora by antibiotics increased the severity of ConA-induced hepatitis, while reconstitution of intestinal flora with H(2)-producing *E. coli*, but not H(2)-deficient mutant *E. coli*, down-regulated the ConA-induced liver inflammation. Furthermore, in vitro production of both TNF-alpha and IFN-gamma by ConA-stimulated spleen lymphocytes was significantly inhibited by the introduction of H(2). These results indicated that H(2) released from intestinal bacteria has biological effects.

References

1. Acott C. A brief history of diving and decompression illness. SPUMS J. 1999;29:98–109.
2. Case E, Haldane JBS. Human physiology under high pressure: I. effects of nitrogen, carbon dioxide, and cold. J Hyg. 1941;41:225–49.
3. Lazarev N, Lyublina E, Madorskaya R. Biological actions of gases under pressure. Leningrad: VMMA; 1941.
4. Zetterstrom A. Deep-sea diving with synthetic gas mixtures. Mil Surg. 1948;103:104–6.
5. Bjurstedt H, Severin G. The prevention of decompression sickness and nitrogen narcosis by the use of hydrogen as a substitute for nitrogen, the Arne Zetterstrom method for deep-sea diving. Mil Surg. 1948;103:107–16.
6. Hengyi Tao, Xuejun Sun et al. Diving medicine. Shanghai: Shanghai Science and Technology Press; 2010.
7. Edel P. Sea-Space Research Co., Inc. Harvey, La 70058 since hydrogen is the lightest of all gases, we might expect it to offer the lowest resistance in a laminar flow system which should promote more rapid diffusion of O_2 and CO_2 within the gas exchange units of the, hydrogen as a diving gas: Proceedings of the Thirty-third Undersea and Hyperbaric Medical Society Workshop. Undersea and Hyperbaric Medical Society; 1987. p. 275.
8. Bennett P, Towse E. The high pressure nervous syndrome during a simulated oxygen-helium dive to 1500 ft. Electroencephalogr Clin Neurophysiol. 1971;31:383–93.
9. Gardette B, Gortan C. Mice and monkeys deep dives in heliox, hydrox and hydreliox gas mixtures-synthesis of COMEX "Hydra" programme, Basic and applied high pressure biology. Rochester: University Press of Rochester; 1994. p. 173–84.
10. Abraini J, Gardette-Chauffour M, Martinez E, Rostain J, Lemaire C. Psychophysiological reactions in humans during an open sea dive to 500 m with a hydrogen-helium-oxygen mixture. J Appl Physiol. 1994;76:1113–3.
11. Gardette B, Gortan C, Delauze HG. Helium in-hydrogen out. A new diving technique. Proceedings of the 23rd Annual Scientific Meeting of the European. Bled: Underwater and Baromedical Society; 1997.
12. Gortan C, Delauze H. Hydra V hydrogen experimental dive to 450 meters, offshore technology conference. Offshore technology conference, 1986.
13. Imbert J, Gortan C, Fructus X, Ciesielski T, Gardette B. Hydra 8: pre-commercial hydrogen diving project, submersible technology: adapting to change. Advances in Underwater Technology, Ocean Science and Offshore Engineering. 1988;14:107–16.
14. Joulia F, Barthèlemy P, Guerrero F, Jammes YT. Wave changes in humans and dogs during experimental dives. J Appl Physiol. 1992;73:1708–12.
15. Gardette B, Massimelli J, Comet M, Gortan C, Delauze H, Comex S, France M. Hydra 10: a 701 msw onshore record dive using "Hydreliox.", XIXth annual meeting of European undersea biomedical society on diving and hyperbaric medicine. Trondheim: SINTEF UNIMED; 1993.
16. Imbert C, Colton J, Long W, Grossman Y, Moore H. A system for saturating in vitro preparations with high pressure O_2, He, H_2, and mixtures. Undersea Biomed Res. 1992;19:49–53.
17. Kayar SR, Miller TL, Wolin MJ, Aukhert EO, Axley MJ, Kiesow LA. Decompression sickness risk in rats by microbial removal of dissolved gas. Am J Physiol-Regul Integr Comp Physiol. 1998;275:R677–82.
18. Kayar S, Fahlman A, Lin W, Whitman W. Increasing activity of H_2-metabolizing microbes lowers decompression sickness risk in pigs during H_2 dives. J Appl Physiol. 2001;91:2713–19.
19. Brauer RW, Way RO. Relative narcotic potencies of hydrogen, helium, nitrogen, and their mixtures. J Appl Physiol. 1970;29:23–31.
20. Brauer R, Way R, Perry T. Narcotic effects of helium and hydrogen and hyperexcitability phenomenon at simulated depths of 1500 to 4000 ft of sea water, toxicity of anesthetics. Baltimore: Williams and Wilkins; 1968. p. 241–55.
21. George M, Craig C, William R, Kevin L, Bernard R. Performance effects with repeated exposure to the diving environment. J Appl Psychol. 1981;66:502.

22. Rostain JC, Balon N. Recent neurochemical basis of inert gas narcosis and pressure effects. Undersea Hyperb Med. 2006;33(3):197–204.
23. Kayar SR, Parker EC, Harabin AL. Metabolism and thermoregulation in guinea pigs in hyperbaric hydrogen: effects of pressure. J Thermal Biol. 1997;22:31–41.
24. Dougherty J Jr. Use of H_2 as an inert gas during diving: pulmonary function during H_2-O_2 breathing at 7.06 ATA. Aviat Space Environ Med. 1976;47:618–26.
25. Dougherty JH Jr, Schaefer KE. Pulmonary functions during saturation-excursion dives breathing air. Aerosp Med. 1968 39(3):289–92.
26. Lenoir P, Jammes Y, Giry P, Rostain J, Burnet H, Tomei C, Roussos C. Electromyographic study of respiratory muscles during human diving at 46 ATA. Undersea Biomed Res. 1990;17:121–37.
27. Ornhagen H, Warkander D, Dahlback G. [abstract] Respiratory mechanics during hydrox breathing at 13 ATM, (1984).
28. Gennser M, Ornhagen H. Effects of hydrostatic pressure, H_2, N_2, and He, on beating frequency of rat atria. Undersea Biomed Res. 1989;16:153–64.
29. Ornhagen H, Adolfson J, Gennser M, Gustavson M, Muren A. [abstract] Performance during hydrogen breathing at 13 atm, (1984).
30. Levitt MD. Production and excretion of hydrogen gas in man. N Engl J Med. 1969;281:122–7.
31. Askevold F. Investigations on the influence of diet on the quantity and composition of intestinal gas in humans. Scand J Clin Lab Invest. 1956;8:87–94.
32. Steggerda F. Gastrointestinal gas following food consumption. Ann N Y Acad Sci. 1968;150:57–66.
33. Levitt MD. Volume and composition of human intestinal gas determined by means of an intestinal washout technic. N Engl J Med. 1971;284:1394–98.
34. Carbonero F, Benefiel AC, Gaskins HR. Contributions of the microbial hydrogen economy to colonic homeostasis. Nat Rev Gastroenterol Hepatol. 2012;9:504–18.
35. Strocchi A, Corazza G, Ellis CJ, Gasbarrini G, Levitt MD. Detection of malabsorption of low doses of carbohydrate: accuracy of various breath H2 criteria. Gastroenterology. 1993;105(5):1404–10.
36. Hammer HF. Colonic hydrogen absorption: quantification of its effect on hydrogen accumulation caused by bacterial fermentation of carbohydrates. Gut. 1993;34:818–22.
37. Gall L. The role of intestinal flora in gas formation. Ann N Y Acad Sci. 1968;150:27–30.
38. Suzuki Y, Sano M, Hayashida K, Ohsawa I, Ohta S, Fukuda K. Are the effects of α-glucosidase inhibitors on cardiovascular events related to elevated levels of hydrogen gas in the gastrointestinal tract? FEBS Lett. 2009;583:2157–59.
39. Nishimura N, Tanabe H, Sasaki Y, Makita Y, Ohata M, Yokoyama S, Asano M, Yamamoto T, Kiriyama S. Pectin and high-amylose maize starch increase caecal hydrogen production and relieve hepatic ischaemia—reperfusion injury in rats. Br J Nutr. 2012;107:485–92.
40. Ito M, Hirayama M, Yamai K, Goto S, Ito M, Ichihara M, Ohno K. Drinking hydrogen water and intermittent hydrogen gas exposure, but not lactulose or continuous hydrogen gas exposure, prevent 6-hydorxydopamine-induced Parkinson's disease in rats. Med Gas Res. 2012;2:15.
41. Kajiya M, Sato K, Silva MJ, Ouhara K, Do PM, Shanmugam K, Kawai T. Hydrogen from intestinal bacteria is protective for Concanavalin A-induced hepatitis. Biochem Biophys Res Commun. 2009;386:316–21.

Chapter 4
Detection Techniques for Hydrogen

Xiao Zhai, Atsunori Nakao and Xuejun Sun

Abstract It is important to observe the pharmacokinetics in vivo for hydrogen bioresearch. Measuring the concentration of hydrogen is one of the key points in this study. Several methods have been developed to detect the concentration, including the gas chromatography technique, the rheophore detection technique, and the oxidimetry technique. The gas chromatography is the most classical, which is capable of quantitatively analyzing the minimum amount and even trace the amount of hydrogen. The rheophore detection measures a large scale from a minimum amount in the tissue to a high amount in the solution. The oxidimetry technique is usually used in the determination of the chemical composition of a hydrogen product. The advantages and disadvantages of these detection techniques are discussed with particular attention on the practical use.

Keywords Hydrogen detection technique · The gas chromatography technique · The rheophore detection technique · The oxidimetry technique

The concentration measurement is the key point in the bioresearch for hydrogen since it is necessary to understand the pharmacokinetics including the absorption regularity, distribution characteristics, and the dose-response relationship [15]. There are a number of methods for hydrogen concentration detection, and detection techniques are almost the same as those for other biological research: the gas chromatography technique is the most classical; the rheophore detection technique is the most convenient; and the oxidimetry technique is the simplest.

The gas chromatography technique can quantitatively analyze the minimum amount and even trace the amount of hydrogen [14]. The rheophore detection

X. Sun (✉)
Department of Navy Aeromedicine, Second Military Medical University, Shanghai, China
e-mail: sunxjk@hotmail.com

X. Zhai
Graduate Management Unit, Changhai hospital affiliated to the Second Military Medical University, Shanghai, China

A. Nakao
Department of Emergency, Disaster and Critical Care Medicine, Hyogo College of Medicine, Nishinomiya, Japan

X. Sun et al. (eds.), *Hydrogen Molecular Biology and Medicine*,
DOI 10.1007/978-94-017-9691-0_4

technique measures a large scale from a minimum amount in the tissue to a high amount in the solution, and it also can continuously detect in in vivo, so its status is unsubstitutive in the dose–response relationship for hydrogen. The oxidimetry technique is usually used in determining the chemical composition of a hydrogen product; however, its demerits include the relatively poor sensitivity and selectivity and the instability affected by various oxidative–reductive substances such as oxygen. Besides, the mass spectrography generally is seldom used but in detection of samples with extremely low concentration of hydrogen.

This chapter concisely introduces these techniques, while the specific manipulation should be complied by the instruments for the equipment and test requirements.

4.1 Gas Chromatography

The gas chromatography technique is extensively used in the determination of the hydrogen concentration and in the solution and blood. For example, Oshawa et al. applied this technique in the first biological article of hydrogen in 2007 [6]. Recently, it was reportedly also used in human blood during and post H_2 inhalation [7,8] (Fig. 4.1).

The gas chromatography technique sets the gas as the mobile phase. It is commonly used for the analysis of impurity in supra-pure gas and light concentration

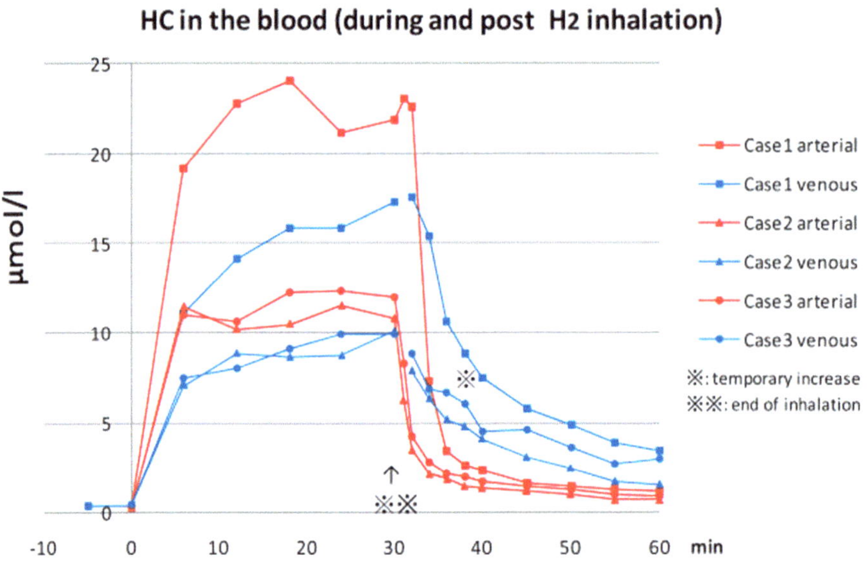

Fig. 4.1 Hydrogen concentration in human blood during and post 1–4 % H_2 inhalation. (From [7])

gas in the industry production. Based on the different immobile phases, it can be classified into two kinds according to the different immobile phases: the gas solid chromatography as the immobile phase is solid and the gas liquid chromatography as the immobile phase is liquid. Based on the principle of segregation, the gas chromatography can also be classified into the adsorption chromatography and the partition chromatography. In the adsorption chromatography, the immobile phase is adsorbent, so the gas solid chromatography belongs to adsorption chromatography while the gas liquid chromatography belongs to the partition chromatography.

Since the gas chromatography is a column chromatography, depending on the thickness of the column used, it can be divided into generally packed columns and capillary columns. Moreover, the capillary column can be divided into open-hole capillary column and packed capillary column. For the open-hole capillary, the stationary liquid is directly applied to the inner wall of the glass or metal capillary with an inner diameter of 0.1–0.5 mm. For the packed capillary column, it was developed in recent years. Certain porous solid particles are filled into the thick glass tube, and then the tube is heated and drawn into the capillary, with an inner diameter of 0.25–0.5 mm. In practice, the gas–liquid chromatography is mainly used in the gas chromatography technique.

Gas chromatography instruments are used for the gas chromatograph. The carrier gas is mostly nitrogen. The column is the packed column or the capillary column. The carrier is celite or porous polymer balls after acid refinement or silanization with a diameter of 0.25–0.18, 0.18–0.15, or 0.15–0.125 mm. The inner diameter of the commonly used glass or elastic quartz capillary is 0.20 or 0.32 mm. The temperature of the inlet should be about 30–50 °C higher than that of the column. The sample size should be less than several microliters, and the thin the column is the least of the sample size.

The gas chromatography system is composed of the adsorbent in the column or the liquid immobile phase applied on the inertia solid, and the gas mobile phase continuously passing the column [10]. The sample is added from the end of the column. Since the immobile phase has a different ability of absorption and dissolution for each component, which means the distribution coefficient differs between the immobile phase and mobile phase, the speed of each component moving forward the column varies. The chromatogram is made with c (the effluent concentration of each component) for the vertical axis and t (time after the sample injection) for the horizontal axis. Then, it is analyzable for the retention time at the maximum concentration, the passing time through the column, and the relationship between the component and the retention time.

The chromatograph peak of the effluent concentration is a curve similar to Gauss distribution rather than a rectangle, because of the zone-broadening phenomenon led by the factors of eddy diffusion, vertical proliferation, and mass transfer resistance during the movement in the column. In the immobile phase of the chromatography column, there are two deposited manners: one is called packing chromatography with the granular absorbent or the inertia solid granulation coating the stationary solid, and the other is called capillary chromatography with the stationary solid coating or chemical cross-linking to the inner wall of the column.

The chromatographic peak signal of the objective constituents detached from the sample is first calculated using the integral equation method, and then is compared to the standard preparation to acquire the concentration of the corresponding components. The gas chromatographic technique is a constituent isolation technique, so hydrogen can be detected after freeing from the solution with the bottle heated. However, the resolution of the concentration differs significantly from different test methods. For example, as the prevalent thermal conductivity detection can only detect the sample with the concentration above several parts per million (ppm); therefore, the hydrogen in the blood, which is at about 0.01 ppm, generally cannot be detected.

Takao et al. [4] made the hydrogen gas analyzer combining the gas chromatographic fractionation technique with the atomic absorption spectrum, featuring high sensitivity. The neon is set as the carrier gas, refined in the depurant cartridge, mixed with the sample gas in the control valve. The mixed gas then arrives at the detached column put in a well-closed container filled with liquid nitrogen (the temperature is below − 196 °C). The carbon monoxide, oxygen, and some other gases are separated from the mixed gas due to its high boiling point, above − 196 °C, while the hydrogen (the boiling point is −252.8 °C) is left and sent to the atomic absorption trace-reducing detector, reductively reacting with the mercuric oxide, and finally changing into water and mercuric vapor. The hydrogen concentration could be calculated by the detected mercuric vapor. This technique is very sensitive, and detects the lowest concentration at 5 ppb. However, the system requires extreme hypothermia, so it is now just used in the laboratory and hard to meet the request in public [3].

4.2 Electrochemistry

To detect the hydrogen concentration, the rheophore detection technique is often used for samples from tissues and the blood. The system produced by Unisense Co., Ltd. from Denmark meets extensive needs of general research. Specifically, the picket age firstly proceeded by standard hydrogen solution at different concentrations to obtain the standard curve, and then electrodes were put in the sample. The measured value will be compared to the standard curve to yield the absolute concentration of the sample (Fig. 4.2).

It features much advantage for hydrogen transducers to detect the concentration of hydrogen especially in the liquid or the tissue. According to the working principle, the hydrogen transducers can be classified into four types: the electrochemistry type, the semiconductor type, the thermoelectricity type, and the optical fiber type. The metal oxide semiconductor has already been industrialized, and the other three types are developing very fast. The following section introduces the specified working principle.

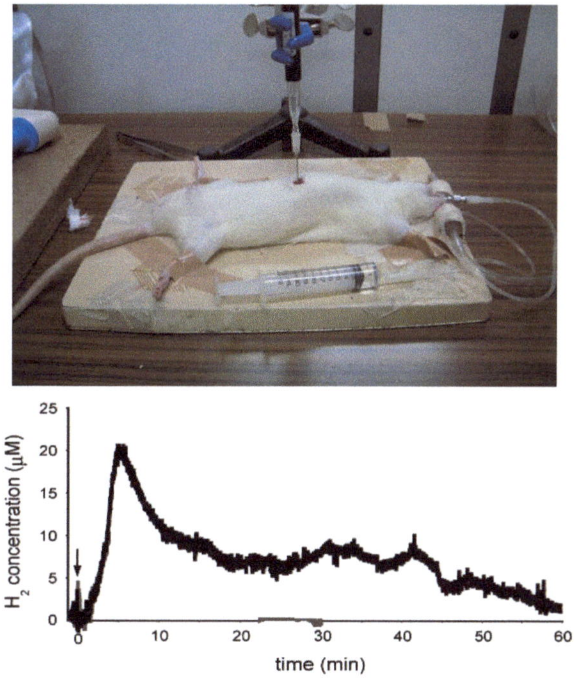

Fig. 4.2 The use of hydrogen electrodes

4.2.1 *Semiconductor Transducer*

The semiconductor transducer includes the electric resistance type and the nonelectric resistance type. The electric resistance type uses tin oxide, zinc oxide, tungsten oxide, and other metallic oxides as the materials for gas absorption [13]. The working principle is that after absorbing by the metallic oxide, hydrogen will be oxidized and release the electron, which will be combined with the oxygen atom and produce the electric current. As a result, the electric current can be detected and the concentration of the hydrogen can be calculated using the electrochemical analysis principle. Besides, the main defect of the system is the poor selectivity since other gases with reducibility, such as carbon monoxide, will interfere with the results. Now some specialists applied nanometer technology and found that it significantly improved the physical and the analysis selectivity. Others added a ratio of better hydrogen selective material such as platinum and also improved the hypersensitivity and the selectivity.

The nonelectric resistance transducer detects the concentration using the capacitance including the Schottky barrier diode and the metal–oxide–semiconductor field-effect transistor. The Schottky barrier diode transducer is a semiconductor material sedimentated with metal. The principle is when hydrogen is exposed to the transducer, it is split into the hydrogen atom, staying at the interface of the semiconductor. With the adscititious offset voltage, the capacitance of the interface changes with the voltage.

In conclusion, the semiconductor-type hydrogen conductor is promising as it is structurally simple, easily integrated, realizably miniaturizated, and constantly serviceable. On the other hand, it is designed to work at a high temperature, increasing the energy consumption and the electric spark, which easily causes explosion in a hydrogen-rich area.

4.2.2 Thermoelectric Hydrogen Sensor

To develop the thermoelectric hydrogen sensor, deposit the thermoelectric material on the basal piece, then apply the hydrogen-catalytic metal such as platinum on a part of the thermoelectric material, and lastly install two electrodes on the metal and the thermoelectric pellicle, respectively.

When the sensitive element is exposed to hydrogen, hydrogen reacts with oxygen and releases heat catalyzed by the metal. So, the side of the thermoelectric material with the metal has high temperature, and the other side without the metal has low temperature. The thermoelectric material transforms the temperature difference into electrical potential. Reading the output electrical signals, the concentration of hydrogen can be acquired [9].

The thermoelectric hydrogen sensor has two advantages: the first is that it does not need an additional auxiliary power unit since the material can transform temperature difference into electric signals directly, and the second is that selectivity of hydrogen can be improved by using a better metal.

4.2.3 Optical Fiber Hydrogen Sensor

The optical fiber hydrogen sensor can be classified into micro-mirror-type sensor and fiber-grating slit-type sensor. Compared to the solid-state hydrogen sensor—the common defect of which is the possibility to generate electric spark since it detects electric signals, which might cause safety problems especially in hydrogen-rich environment—the optical fiber sensor using the optical signal is more suitable for explosive environment. The optical fiber hydrogen sensor mostly adopts the metal palladium and palladium alloys as hydrogen gas-sensing material, featuring fine selectivity. Optical fiber hydrogen sensor technology detects the concentration of hydrogen by measuring the film transmittance, reflectance, and other physical parameters [11].

To get the fiber-grating slit-type hydrogen-sensitive element, a fiber Bragg grating is plated with a layer of palladium. The working principle is when this element is placed in a hydrogen atmosphere, it uptakes the hydrogen and forms palladium hydride. Then, the deformation of the palladium membrane structure occurs, causing the change of grating wavelength. Lastly, the concentration of hydrogen is detected by measuring the change of grating wavelength.

To obtain the micro-mirror-type optical fiber hydrogen sensor, one end of the optical fiber of the micro-mirror-type hydrogen sensor is coated with a layer of palladium or palladium alloy membrane [16]. When the incident light reflects at the sensitive element and enters into the light detector, the reflectivity of the palladium membrane changes after the absorption of hydrogen, causing the signal to change. So, the concentration of hydrogen is identified [5]. This kind of sensor has the advantages of simple principle and most mature development at present and more feasible in a relatively high concentration of hydrogen environment.

It is the main reason that problems of selectivity, security, stability, sensitivity, and the weakness of output signals and other issues limit the applications of the hydrogen sensor. These problems have been gradually resolved, especially with the development of an optical fiber sensor [1], nano [2] and novel sensitive materials [17], and so on, to improve the hydrogen sensor technology.

4.3 Redox Titration

4.3.1 A Simple Method for Testing Hydrogen Content in Solution

The determination accuracy of the concentration of hydrogen is a very important problem facing both scientific research and practical application of hydrogen. The standard method for the determination of the concentration of hydrogen is the gas chromatography and electrode method, but these two methods need expensive equipments and complex detection technology, which are not suitable for some small laboratory and normal users. So, there is a necessity to establish a simple and accurate method of hydrogen detection. The MiZ company in Kanagawa Prefecture, Japan, which specializes in hydrogen-related products, recently established a simple method for the detection of hydrogen concentration.

Bunpei Sato [12], a researcher in the company, published the article recently in the *Journal of Medical Gas Research,* and introduced the economical method for the determination of the hydrogen content in solution (the company has developed a hydrogen injection preparation technology without opening the package). The method belongs to the classical oxidation–reduction titration, and the principle is that catalyzed by colloidal platinum (Pt), hydrogen reduces methylene blue (MB). MB is a kind of commonly used dye and redox titration indicator (Fig. 4.3), which was used in the very famous scientific experiment of redox reaction of glucose and MB. Because of the influence of the covalent bonds of hydrogen itself, hydrogen is difficult to reach with MB at room temperature. But by the catalytic agent of colloidal platinum, the molecular hydrogen reduces the oxidized blue MB into colorless reduced MB (leucoMB, leucomethylene blue; Fig. 4.4). The chemical reaction formula is as follows:

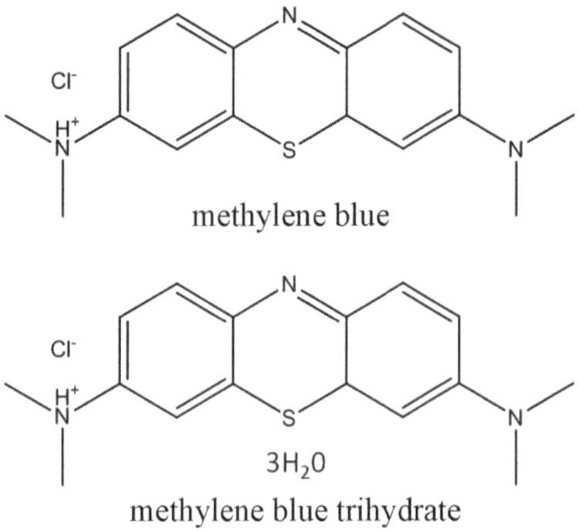

Fig. 4.3 Molecular structure of methylene blue and methylene blue trihydrate

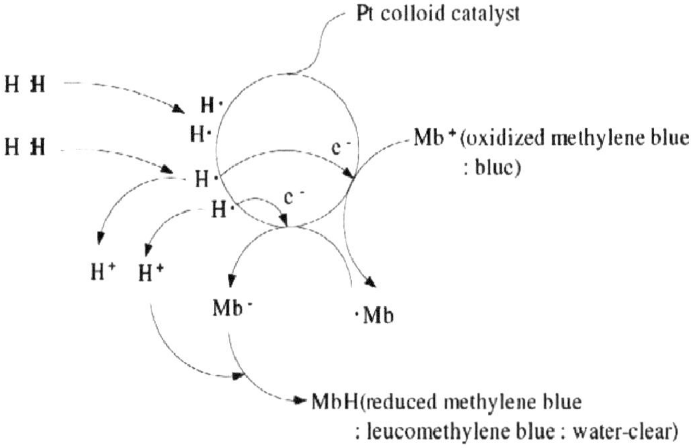

Fig. 4.4 By the catalytic agent of colloidal platinum, 1 mol molecular hydrogen reduces 1 mol oxidized blue methylene blue into 1 mol colorless reduced methylene blue. (From [12])

$$\mathrm{MB\,(blue)} + 2\mathrm{H}^{+} + 2\mathrm{e}^{-} = \mathrm{leucoMB\ (colorless)}.$$

According to the above formula, the authors use the redox titration volumetric analysis method to determine the concentration of hydrogen in the solution.

4.3.2 The Experimental Method

4.3.2.1 Configuration of Methylene Blue Platinum Reagent

The MB is purchased from Waldeck-Gmbh & Co Waldeck KG, Munster, Germany. The 2 % colloidal platinum solution is purchased from the Tanaka Kikinzoku Group Company. 0.3 g MB is dissolved in 98 % ethanol (98.9 g) solution to get 99.2 g MB ethanol solution, and then 0.8 g colloidal platinum solution is mixed with 99.2 g MB ethanol solution into 100 g MB and colloidal platinum MB-Pt reagent (MiZ Company, Kanagawa, Japan). The MB-Pt reagent is packaged in a small plastic bottle, which is assembled with a straw (17 mg or 0.02 ml per drop).

4.3.2.2 Preparation of Aqueous Solutions of Hydrogen

Hydrogen-saturated water (0.8 mM) is prepared by pure water bubbling method, and then diluted in accordance with the proportion of pure water to 0.3, 0.2, and 0.1 mM hydrogen water solution (not specific). The concentration of hydrogen is detected by the method of electrochemical detection, using the electrochemical gas.

The redox titration method is used to detect the hydrogen content. The MB-Pt is titrated in 20 ml hydrogen solution until the color of blue is no longer disappeared. Then the concentration is determined and calculated (Fig. 4.5).

4.3.3 Results

Since a drop of solution is supposed to contain 17 mg, 0.16 μmol Mb, 100 drops of consumed solution is believed to contain 20 ml saturated hydrogen water (0.8 mM) (1.7g; per ml of saturated hydrogen water consumes five tutration solution). When the titrant instills in the hydrogen solution, the number of moles of hydrogen can be represented by the MB, the following formula can be used:

$$\text{Hydrogen mole number} = \text{titration droplet number} \times (17/1000) \times (0.3/100)/319.85$$

$$800\,\mu\text{mol}/(1000\,\text{ml}/20\,\text{ml}) = 16\,\mu\text{mol}$$

$$16\,\mu\text{mol}/0.16\,\mu\text{mol/drop} = 100\ \text{drops}.$$

Although the theoretical calculation is 100 drops, the actual testing found that 55 drops can reach the titration point. (Note: the author explained that hydrogen can be volatile, but it seems unlikely to reach 45 % in such a short time. One possibility is that the solubility of hydrogen is not probably 0.5 mM instead of the so-called 0.8 mM, or the ventilated standard liquid did not reach 0.8 mM. The possibility of calibration error is present.) According to the research results, the authors thought

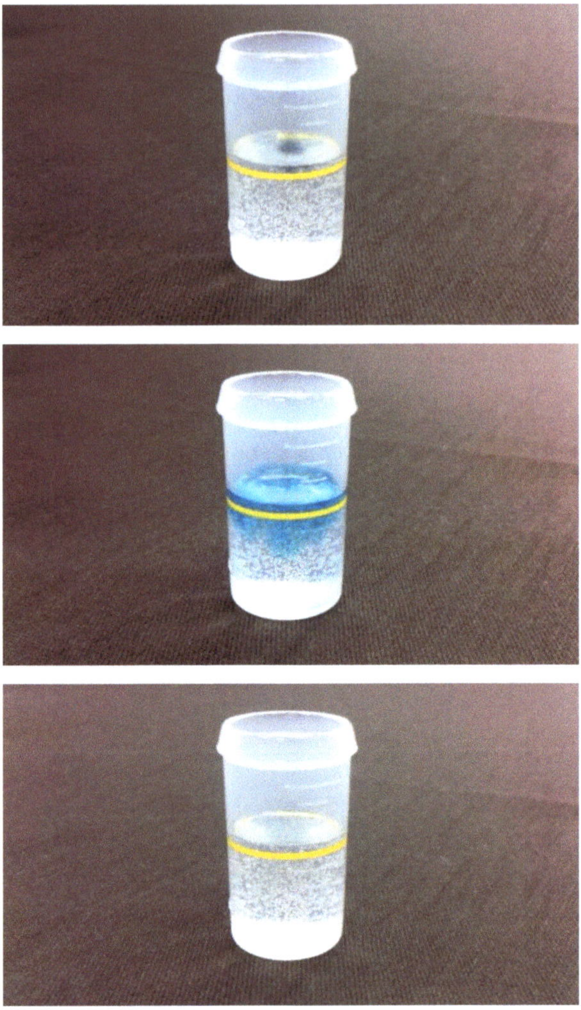

Fig. 4.5 Titration results. The first bottle shows for just adding the titration solution, the second bottle after adding the titration solution, and the third bottle for 10 s after the titration solution. (From [12])

that each drop can neutralize 0.29 μmol hydrogen (16 μmol/55). That means each drop can consume 20 ml hydrogen in the solution (14.5 μmol/L, 0.03 mg/L, 0.3 ppb, or 0.0003 ppm).

$$0.29\,\mu\text{mol/drop} \times (1000\,\text{mL}/20\,\text{mL}) = 14.5\,\mu\text{mol/L (or } 0.03\,\text{mg/L)}$$

The results of the analysis showed that as the water used for dilution contained oxygen, which would perform oxidation with the hydrogen-reduced MB, the results would be affected with distortion. So, the results using nitrogen-saturated water as diluent might be more stable, and more accordant with the expected results. The linear analysis found that the solution oxygen can cause certain interference, but the relative proportion was small and did not affect the final results, so this method can be used as alternative electrode detection method.

However, this titration method has obvious limits. This method is used for the detection of pure water, not containing other redox components. The accuracy would be interfered if there are other oxidizing or reducing substances. For example, the blood and the cell culture are not suitable for this method. From the practical point of view, this method is ideal for qualitative not quantitative analysis.

References

1. Barmenkov YO, Ortigosa-Blanch A, Diez A, Cruz JL, Andres MV. Time-domain fiber laser hydrogen sensor. Opt Lett. 2004;29(21):2461–3.
2. Huang BR, Yang YK, Cheng HL. Rice-straw-like structure of silicon nanowire arrays for a hydrogen gas sensor. Nanotechnology. 2013;24(47):475–502. doi:10.1088/0957-4484/24/47/475502.
3. Kaal E, Janssen HG. Extending the molecular application range of gas chromatography. J Chromatogr A. 2008;1184(1–2):43–60. doi:10.1016/j.chroma.2007.11.114.
4. Kawano T, Tsuboi N, Tsujii H, Sugiyama T, Asakura Y, Uda T. Stability test and improvement of hydrogen analyzer with trace reduction detector. J Chromatogr A. 2004;1023(1):123–7.
5. Mandelis A, Garcia J. Pd/PVDF thin film hydrogen sensor based on laser-amplitude-modulated optical-transmittance: dependence on H_2 concentration and device physics. Sensors and Actuators B Chemical 1998;49(3):258–267. doi:10.1016/S0925-4005(98)00137-3.
6. Ohsawa I, Ishikawa M, Takahashi K, Watanabe M, Nishimaki K, Yamagata K, Katsura K, Katayama Y, Asoh S, Ohta S. Hydrogen acts as a therapeutic antioxidant by selectively reducing cytotoxic oxygen radicals. Nat Med. 2007;13(6):688–94. doi:10.1038/nm1577.
7. Ono H, Nishijima Y, Adachi N, Sakamoto M, Kudo Y, Kaneko K, Nakao A, Imaoka T. A basic study on molecular hydrogen (H2) inhalation in acute cerebral ischemia patients for safety check with physiological parameters and measurement of blood H2 level. Med Gas Res. 2012;2(1):21. doi:10.1186/2045-9912-2–21.
8. Ono H, Nishijima Y, Adachi N, Sakamoto M, Kudo Y, Nakazawa J, Kaneko K, Nakao A. Hydrogen (H2) treatment for acute erythematous skin diseases. A report of 4 patients with safety data and a non-controlled feasibility study with H2 concentration measurement on two volunteers. Med Gas Res. 2012;2(1):14. doi:10.1186/2045-9912-2–14.
9. Park SC, Yoon SI, Lee CI, Kim YJ, Song S. A micro-thermoelectric gas sensor for detection of hydrogen and atomic oxygen. Analyst. 2009;134(2):236–242. doi:10.1039/b807882c.
10. Purcaro G, Moret S, Conte L. Hyphenated liquid chromatography-gas chromatography technique: recent evolution and applications. J Chromatogr A. 2012;1255:100–11. doi:10.1016/j.chroma.2012.02.018.
11. Sekimoto S, Okazaki S, Fukuda K, Asakura S, Shigemori S, Takahashi T, Nakagawa H. A fiber-optic evanescent-wave hydrogen gas sensor using palladium-supported tungsten oxide. Sensors and Actuators B Chemical 2000;66(1/3):142–145. doi:10.1016/S0925-4005(00)00330-0.
12. Seo T, Kurokawa R, Sato B. A convenient method for determining the concentration of hydrogen in water: use of methylene blue with colloidal platinum. Med Gas Res. 2012;2:1. doi:10.1186/2045-9912-2–1.
13. Shukla S, Ludwig L, Cho HJ, Duarte J, Seal S. Effect of air-pressure on room temperature hydrogen sensing characteristics of nanocrystalline doped tin oxide MEMS-based sensor. J Nanosci Nanotechnol. 2005;5(11):1864–74.
14. Solomons NW, Viteri FE, Hamilton LH. Application of a simple gas chromatographic technique for measuring breath hydrogen. J Lab Clin Med. 1977;90(5):856–62.

15. Spohr A, Guilford WG, Haslett SJ, Vibe-Petersen G. Use of breath hydrogen testing to detect experimentally induced disaccharide malabsorption in healthy adult dogs. Am J Vet Res. 1999;60(7):836–40.
16. Wei X, Wei T, Xiao H, Lin Y. Nano-structured Pd-long period fiber gratings integrated optical sensor for hydrogen detection. Sensors and Actuators B Chemical 2008;134(2):687–93. doi:10.1016/j.snb.2008.06.018.
17. Xu L, Du J, Chen B. Preparation and electrocatalytic activity of nanocrystalline Ni-Mo-Co alloy electrode for hydrogen evolution. J Nanosci Nanotechnol. 2013;13(3):2016–20.

Chapter 5
Selective Antioxidative Effect of Hydrogen

Qiang Sun, Wenjie Han and Atsunori Nakao

Abstract Selective antioxidation is the mechanism underlying the biological effect of hydrogen, which has been widely recognized. We cannot fully understand this mechanism until we are familiar with radicals, reactive oxygen species, and oxidative damage. In order to let the readers have a general understanding of free radical biology, this chapter provides knowledge which is closely related to the selective antioxidation of hydrogen. If someone wants to get a more comprehensive understanding of the free radical biology, references related to free radicals will be needed.

People believe that free radicals or reactive oxygen species is a main source of sickness, which exaggerates the negative effects of free radicals or reactive oxygen species. In fact, the oxidative stress is crucial in maintaining the body's normal function. There exist different types of free radicals or reactive oxygen species, most of which are beneficial to the body; only a small number of them that are highly reactive are key to the oxidative damage. Selective antioxidation and endogenous antioxidant are the most reliable means to resist oxidative damage. The finding of selective antioxidation and endogenous antioxidant of hydrogen will be two of the important achievements in the field of free radical biology.

Keywords Reactive oxygen species (ROS) · Free radicals · Hydrogen · Selective antioxidation · Oxidative stress

A. Nakao (✉)
Department of Emergency, Disaster and Critical Care Medicine, Hyogo College of Medicine, Nishinomiya, Japan
e-mail: atsunorinakao@aol.com

Q. Sun
Department of Hyperbaric Medicine, Navy General Hospital, Fucheng Road, Beijing, China

W. Han
Department of Cadre Ward, Navy General Hospital, Fucheng Road, Beijing, China

X. Sun et al. (eds.), *Hydrogen Molecular Biology and Medicine*,
DOI 10.1007/978-94-017-9691-0_5

5.1 Reactive Oxygen Species Have Many Biological Effects

Free radical is an atom or group of atoms that has an electron that is not part of a pair, causing it to take part easily in chemical reactions. The oxygen molecule is also a kind of free radical which has two unpaired electrons. Because spins of the electrons are parallel, this molecule is stable. According to the above definition of free radical, many metal ions which have an important biological effect such as iron and copper ion also belong to the family of free radicals because of unpaired electrons. Reactive oxygen species (ROS) are chemically reactive molecules containing oxygen which are formed as a natural by-product of the normal metabolism of oxygen. Common ROS in the body include oxygen ion, superoxide anion radical, hydrogen peroxide, hydroxyl radical, nitric oxide, nitrous acid anion, and so on.

5.1.1 Oxygen Is One Type of Free Radical

The vast majority of life on earth, including humans, need oxygen, without which life cannot survive. Since oxygen itself is one type of free radical, from this perspective, at least this free radical is important to life and health. Some people may think oxygen is a special type of free radical whose toxicity and hazard is less than those of more reactive free radicals. Some people even believe that all the free radicals are harmful to the body except oxygen. On the contrary, many free radicals are important bioactivators like oxygen [1].

Cells are the body's most basic structural and functional units; high-level organisms are complex systems composed of numerous cells, such as the adult body, made of about 160 billion cells. To maintain normal cell function, oxygen is continuously used to generate the metabolic energy substance ATP (adenosine triphosphate). ATP is a nucleoside, often called the "molecular unit of currency" of intracellular energy transfer [2]. ATP transports chemical energy within cells for metabolism.

Probably a lot of people do not know specifically why oxygen is very important to aerobic biological terms. The reason why oxygen is important and indispensable is that oxygen is the only final electron acceptor in the body and no other substance can replace it. From a chemical point of view, the electron acceptor can be thought to have an oxidation effect. There are many oxidizing substance like ROS in human body which are capable of receiving electrons, the ability of which originates from oxygen. Without oxygen, cells will lose the ability to receive electrons. Electrons in the human body are often in the form of hydrogen atom, the source of which is from the uptake of glucose (e.g., via the Krebs cycle), protein, fat, etc. Under the catalysis of the various enzymes, generated hydrogen enters into the mitochondria in the form of NADH and FADH2. Through oxidative phosphorylation in this electron transfer process, the last four electrons are received by the oxygen molecules, which then produce water. Notably, although the Krebs cycle and oxidative phosphorylation is composed of many redox reactions, oxygen is involved in the final stage of almost all of the processes [3].

From a macro perspective, cellular energy metabolism can be simplified as the reaction between oxygen and hydrogen atoms which produces water; most of the energy metabolism in the human body is ready for this process. This is the basic way of energy metabolism; energy metabolism cannot continue in the cells without oxygen. In the evolution of life, with the appearance of mitochondria (an important subcellular structure), the cells obtained an important capability that under normal temperature conditions (e.g., 37 °C in many animals), with the help of a series of enzymes, they can degrade many energy substances into donor electrons which may be supplied directly to oxygen, meanwhile generating energy and ATP.

The most important way for aerobic cells to produce ATP is oxidative phosphorylation, which is the electron transfer process in the mitochondria. It should be known that the electron transfer is equal to the redox reaction; there are many free radicals which are involved in the oxidative phosphorylation process, e.g., ubiquinone [4]. Radical reactions are of great importance for cells to produce energy, without which cellular energy supply will be difficult to be continued as well as cell function. Therefore, the important roles free radicals play in maintaining cell function cannot be replaced.

5.1.2 Nitric Oxide Is One Type of ROS Which Has Important Physiological Functions

The most famous ROS which can be thought of as a signaling molecule is nitric oxide. Nitric oxide, which has unpaired electrons, is a typical free radical and also a very important gaseous signaling molecule. In 1980, US scientist Furchgott's study found a small molecule which played a role in vascular smooth muscle relaxation, later named vascular endothelial relaxing factor (endothelium-derived relaxing factor, EDRF), an unstable free radical [5]. EDRF was then identified as nitric oxide. In 1987, Moncada et al. observed a relaxant effect of EDRF on vascular smooth muscle; meanwhile, they confirmed nitric oxide released by endothelial cells through the chemical method. He explained the extent of its relaxant effect on vascular smooth muscle according to its concentration [6]. In 1988, Palmer et al. proved that L-arginine is a precursor to nitric oxide synthesized by vascular endothelial cells; this established the concept that nitric oxide can be synthesized in mammals [7].

As the most important regulatory factor of blood vessels, when vascular endothelial cells send orders to relax the muscles and promote blood circulation, nitric oxide molecules will be generated, which is small enough and can easily pass through cell membranes. Vascular smooth muscle cells relax and mediate blood vessel vasodilatation after receiving the relaxation signal. As we all know, nitroglycerin is the drug which treats angina pectoris; people have been trying to clarify its therapeutic mechanism at the molecular level over the years. Studies found that nitroglycerin and other organic nitrates themselves have no effect; they are first converted to nitric oxide in vivo which stimulates cGMP formation within the vascular smooth muscle and then relaxes blood vessels. This effect is similar to that of EDRF [8].

Studies on the effect of nitric oxide on the central nervous system suggest that nitric oxide could activate adjacent peripheral neurons such as presynaptic nerve terminals and astrocyte by diffusion, and then activate guanylate cyclase to improve cGMP levels [9]. Nitric oxide can induce learning and memory-related long-term potentiation (LTP), which plays a role as a retrograde messenger in LTP. Long-term depression (LTD) occurs at synapses in cerebellar Purkinje neurons, which receive two forms of excitatory input, one from a single climbing fiber and one from hundreds of thousands of parallel fibers. Cerebellar LTD is thought to lead to motor learning in which nitric oxide is involved [10]. Nitric oxide also exists in the peripheral nervous system. Nitric oxide is considered as a noncholinergic and nonadrenergic nerve neurotransmitter or media which is involved in the process of pain afferents and feeling delivery. According to reports, nitric oxide plays an important intermediary role in the gastrointestinal-nerve-mediated gastrointestinal smooth muscle relaxation. In the gastrointestinal plexus, the coexistence of nitric oxide synthase (NOS) and vasoactive intestinal peptide could cause nonadrenergic noncholinergic (NANC) relaxation [11]. A vasoactive intestinal peptide antibody could only partially eliminate the NANC relaxation, the rest of which could be eliminated by N-methyl-arginine. As a NANC neurotransmitter, neuronal nitric oxide plays an important role in the genitourinary system, including physiological functions such as urination control, which provides a theoretical basis for the drug treatment of genitourinary system diseases [12]. It has been demonstrated that there is a wide range of nitric oxide (NO)-mediated nervous system in the human body, which is as important as the cholinergic nerves, adrenaline nerves, and peptides. When the brain sends a message via peripheral nerves, the perineal vascular will be supplied with nitric oxide, which can dilate blood vessels, increase blood flow, and thereby enhance erectile function. In some cases, erectile dysfunction is caused by low nitric oxide production by the nerve endings. "Viagra" can expand the effectiveness of nitric oxide and thereby enhance erectile function.

Nitric oxide has a close relationship with the immune function. When toxins or T cells activate macrophages and polymorphonuclear leukocytes, large amounts of inducible NOS and superoxide anion will be produced, which will synthesize large amounts of nitrous oxide nitrogen. The reaction of nitric oxide and superoxide anion can directly produce nitrite anion, which plays a very important anti-inflammatory role of killing bacteria, fungi, tumor cells, etc. [13].

5.1.3 Other Species of ROS

In addition to nitric oxide, more and more scientific evidence suggests that many ROS, e.g., superoxide anion, hydrogen peroxide, nitrous acid anions, etc., can act as intracellular signaling molecules which regulate many cellular functions.

Ninety-eight percent oxygen of the human body is used to produce energy by oxidative phosphorylation. Only about 2 % oxygen receive a single electron to produce superoxide anion. There are many ways for cells to produce superoxide anion, but most of them are produced by being partially deoxygenized. Superoxide anion is a water-soluble, low-liposoluble ion, which cannot easily gain entrance the cell membrane but can accumulate in a local area. In order to avoid producing a high concentration of superoxide anion which may influence the biological activity of molecule function, the cell evolved an effective enzyme superoxide anion, which is called superoxide dismutase (SOD). We are familiar with SOD; there exist many kinds of SOD which are distributed in the cytoplasm, mitochondria, and outside of the cell. SOD can quickly convert superoxide anion into hydrogen peroxide [14].

Because of the presence of SOD, the intracellular concentration of hydrogen peroxide is 1000 times more than the superoxide anion. Since hydrogen peroxide is liposoluble and it can easily gain entrance into the cell membrane, it can diffuse quickly among cells, which is an important characteristic of signaling molecules. In fact, hydrogen peroxide is recognized as an important ROS-signaling molecule. There are three or more enzymes which are involved in the clearance of intracellular hydrogen peroxide, such as glutathione peroxidase, catalase, and other antioxidant enzymes. Glutathione peroxidase needs glutathione, but catalase and peroxidase can directly reduce hydrogen peroxide to water.

How does hydrogen peroxide play the role of a signal molecule? First, ROS such as hydrogen peroxide are chemically active, which could easily cause a redox reaction with the target molecule. In fact, the signal is transmitted by a redox modification of the target molecule. Experiments showed that signal could be transmitted by ROS through redox activate of mercapto center of the target molecule. Secondly, ROS such as hydrogen peroxide can regulate redox signal transduction by changing the overall level of glutathione and the ratio of oxidized glutathione and glutathione.

Protein phosphorylation is a posttranslational modification of proteins in which a serine, a threonine, or a tyrosine residue is phosphorylated by a protein kinase by the addition of a covalently bound phosphate group. Its reversal process is catalyzed by a protein phosphatase, which is called protein dephosphorylation. ROS such as hydrogen peroxide often cause some changes in the activity of kinase or protein phosphatase in the cell, and then stimulate a series of signal transmission by phosphorylation and dephosphorylation reactions. Signal transduction occurs when an extracellular signaling molecule triggers a biochemical chain of events inside the cell, creating a response which alters the cell's metabolism, shape, and gene expression. ROS such as hydrogen peroxide can activate the transcriptional factor. Cytoplasmic free calcium concentration is closely related to various biological effects. Cytoplasmic calcium concentration depends on calcium pump and calcium channel activity or openness in the cell membrane, endoplasmic reticulum, and mitochondria. IP3 receptors and ryanodine receptor calcium and sodium exchanger, which are related to the calcium channel of the endoplasmic reticulum, are subject to redox regulation. ROS such as hydrogen peroxide are still a hot research topic at present, though many details still need to be confirmed [15].

5.1.4 ROS and Immune Function

As early as 40 years ago, it was recognized that the respiratory burst of inflammatory cells can produce large amounts of ROS [16]. It was mistakenly thought that this is the only positive effect of ROS. The most important biological effect of respiratory burst is that inflammatory cells can use the cytotoxic activity of ROS to directly kill external microorganisms such as bacteria and viruses. Later findings showed that respiratory burst had a more complex meaning; besides killing external microorganisms, it also plays an important role in dealing with damage to the body and macromolecules [17].

Not only are free radicals or ROS such as oxygen, superoxide anion, nitric oxide, etc., ordinary biologically active molecules but they also have an important role in the basic life processes including energy metabolism, signal transduction, and immune function. Therefore, ROS are beneficial to the body's functions that maintain the health of an organism.

5.2 Oxidative Stress and Oxidative Injury

Free radicals and ROS are important biologically active substances, whether the presence of oxidative damage exists or not. In fact, in biological systems, oxidative damage does exist, and many related diseases are caused by free radicals and ROS. The substance that has important physiological functions can also cause damage, is this a contradiction?

5.2.1 Balance Is Important for Life

Everything has two sides. For example, oxygen is very important; hypoxia occurs when the concentration of oxygen in the air is less than 15 %. However, when oxygen concentration is more than 70 %, breathing oxygen for a long time will cause serious damage to the lungs. We cannot survive for more than 1 month if the ambient air is composed of pure oxygen. Since oxygen itself belongs to the family of free radicals, high oxygen concentration and long exposure time will cause severe pulmonary oxidative damage, which is called chronic oxygen toxicity. Breathing oxygen under high pressure can lead to damage to the whole body, especially brain function, manifested as grand mal seizures in severe cases, which is called acute oxygen toxicity. Another example is the blood glucose without which we cannot survive [18]. If the blood glucose concentration is too low (lower than 4.0 mM), hypoglycemia will occur. Hypoglycemia is more dangerous than hyperglycemia, easily threatening the lives of patients. People are familiar with the hazards of hyperglycemia, but the severity of hypoglycemia often lacks attention [19]. Another example is trace elements such as calcium. Most of us are worried about calcium deficiency, but we may not know that an increase in intracellular free calcium concentration is an important mediator of almost all cell damage and even death [20]. The same phenomenon can be seen in balance of acid and base, hormone levels, nerve-excitatory amino acids, hypertension and hypotension, hypothermia and hyperthermia, and so on. Such examples abound in biological systems. It can be said maintaining the biological function means keeping a dynamic balance of a variety of factors. Many important substances and life indicators must be maintained within a limited range, because too low or too high concentrations are unable to maintain normal physiological processes. This phenomenon is known as homeostasis in physiology.

The above examples illustrate whether a biologically active substance is harmful depending on the specific conditions; it is useful in some cases but may be harmful in other cases. We cannot simply consider ROS as the source of sickness. Many commercial promotions exaggerate the beneficial effects of their products freely, which is obviously wrong.

5.2.2 Oxidative Stress

It has gone through several important stages since free radicals were first found by the field of medical biology. In 1775, the British chemist Priestley discovered oxygen by heating mercuric oxide; people came to know oxygen is necessary for the body [21]. After that, people came to realize the toxic effects of oxygen. It was the French scientist Paul Bert who first found acute toxic effects of oxygen in 1887, and then in 1891 chronic pulmonary oxygen toxicity was found [22].

Before 1900, people's understanding of free radicals was very limited; they theoretically speculated a class of active molecules in some of the chemical reaction processes. In 1900, Gomberg successfully prepared the trityl radical; people came to recognize the presence of free radical alone. This research caused the birth of free radical chemistry, but at that time free radicals were not thought to have a role in biology. Until 1950–1960, radiation biology researchers found that radiation damage could increase with radiation-induced free radicals, which launched free radical biology [23]. Obviously, at this stage, free radicals were only thought as a medium in the radiation damage; no one realized the presence of free radicals in the body, but the concept of free radicals being factors in the process of tissue cell damage was widely accepted.

In 1969, Duke University scholars McKord and Fridovich extracted SOD in red blood cells of cattle, and proved the in vivo role of SOD as a catalyst in a disproportionation reaction which can produce hydrogen peroxide from superoxide anion [24]. This was found to be of great significance, because it made people realize that free radicals are a normal component of the body. However, in the past, radiation biology researchers ignored the positive effect of free radicals. In 1973, Babior et al. proved that neutrophils will produce many ROS which can kill bacteria after being stimulated by bacteria; this research first recognized the positive effects of free radicals [25]. But people still believe that this is the only benefit of free radicals in the cell; its toxic effects on the body are still its major role. In 1981, Granger et al. proved the relationship between ischemia–reperfusion injury and free radicals, which led to a burst of free radical research [26]. Obviously, since ischemia–reperfusion injury is an important pathophysiological phenomenon involving many clinical courses, this finding strengthened the traditional understanding of toxic free radicals.

Free radicals are the cause of ischemic injury and tissue inflammation, which are main causes of diseases. At this time, the main point of view was that radical-scavenging substances might treat various types of damage caused by free radicals. However, later studies showed this view to be overly optimistic.

The concept of oxidative stress originated from aging. In 1955, the British scholar Harmna first proposed the free radical theory of aging, in which free radicals were considered as the root of aging by causing tissue damage of biomacromolecules and malignant diseases such as cancer [27]. In 1990, the American professor Sohal, who is an authority on aging research pointed out shortcomings in the free radical theory and first proposed the concept of oxidative stress [28].

Sometimes, we call reactive nitrogen as ROS which include superoxide anion, hydroxyl radicals and hydrogen peroxide, nitrogen dioxide and peroxynitrite, and so on. There are two types of antioxidant systems in the body: one is the antioxidant enzyme system, including superoxide SOD, catalase (CAT), glutathione peroxidase (GSH-Px), etc.; the other is the nonenzymatic antioxidant system, including vitamin C, vitamin E, glutathione, melatonin, α-lipoic acid, carotenoids, trace elements such as copper, zinc, selenium (Se), and so on. Nowadays, oxidative stress reflects an imbalance between the systemic manifestation of ROS and a biological system's ability to readily detoxify the reactive intermediates or to repair the resulting damage.

The main view of oxidative stress is that most of the health problems associated with aging, such as wrinkles, heart disease, and Alzheimer's disease, are caused by too much oxidative stress. This view, which is one-sided, appeared in the early stages of free radical study. Dr. Dunham from the University of California, Berkeley, pointed out that few people could live up to their maximum life span. People usually died of various diseases very young, most of which were caused by free radicals. This suggested that free radicals are important causes of premature human aging; a fight against free radicals could slow down aging and extend life span.

From a certain perspective, almost all human organs are indeed subject to the damage caused by oxidative stress which will be manifested as numerous symptoms, such as fatigue, weakness, muscle and joint pain, indigestion, anxiety, depression, pruritus, headaches, inattention, difficult-to-cure infection, and so on. Most common diseases caused by elevated levels of oxidative stress include heart disease, cancer, osteoarthritis, rheumatoid arthritis, diabetes, and neurodegenerative issues such as Alzheimer's disease and Parkinson's disease.

If the above views of oxidative stress are correct, then the damage caused by oxidative stress can be easily controlled by increasing the body's antioxidant capacity; antioxidants could have very good treatment and prevention effects on the above diseases and aging as well. But unfortunately, no expected effects were found in subsequent experiments; neither antioxidant vitamins, such as vitamin A, C, and E nor some natural antioxidants were proven to treat or alleviate these diseases.

Recently, the view was updated in the field of antioxidants: the reason why an antioxidant is of no effect is that the oxidation and antioxidant systems exist as a network in the living body. In order to effectively improve the body's antioxidant capacity, all levels of it should be upregulated through using a variety of antioxidants. Clearly, the means of combination using of antioxidants is simply to optimize the antioxidant effect, which lead to little concept innovation of the traditional view of antioxidant.

5.2.3 The Essence of Oxidative Damage

Although in the past no clinical effect was found by means of antioxidation, the relationship between oxidative damage and various diseases is gradually being accepted. The academics get a consensus that oxidative damage is indeed a very important issue and the effect of antioxidant therapy worth trying. This thought implies that the academics don't think the effect of antioxidant therapy is promising.

The above chapter has partially answered the inefficiency of antioxidant therapy, because the free radicals and ROS are important bioactivators. Although oxidative damage is caused by an increase in free radicals and ROS, excess antioxidation is not a proper way of treating oxidative damage. Because of the functional importance of free radicals and ROS, they must be maintained in a certain concentration. Emphasizing their toxic effects while ignoring their physiological role will inevitably lead to overemphasizing the effect of antioxidants, which is currently the main problem in antioxidant therapy.

So what is the essence of oxidative stress? We cannot understand oxidative stress until we fully understand the essence of antioxidation; then what is the essence of cellular antioxidation? We have already discussed that oxygen is the only terminal electron acceptor—the real source of oxidation *in the human body*. The essence of antioxidation is to "clear" the oxygen, but the cells happen to get the energy in this process simultaneously by evolution.

Let us briefly review the evolution of cells as early primitive life forms on earth. There was little or no oxygen. At that time, most of the original cells were anaerobic bacteria, which cannot tolerate the toxicity of oxygen, so only bacteria with very simple structures could survive. Later, a class of photosynthetic prokaryotes (cyanobacteria) appeared in large numbers, resulting in the continuous rising of the concentration of oxygen in the atmosphere, which prompted primitive oxygen-consuming bacteria to appear. In the past, most of the anaerobic bacteria could not survive in oxygenated conditions; the oxygen-consuming bacteria were later swallowed by one kind of anaerobic bacteria, and then these anaerobic bacteria obtained the ability of oxygen consumption and thereby could survive under oxygenated conditions. This bacterium is the prototype of mitochondria in cells. Therefore, mitochondria are the most important cellular structures that work against oxidation. The mitochondria utilizes 89% oxygen, which can be thought to be the main means for cells to confront oxygen toxicity; this is the most successful evolutionary outcome of the cell. Mitochondrial antioxidant capacity is associated with the electrons provided by sugar, protein, and fat.

If 2% oxygen in the cells will be converted into ROS, then where do the ROS-removing electrons come from? The answer is still derived from electrons provided by sugar, protein, and fat. For example, a reducing agent such as vitamin C and vitamin E will be oxidized after scavenging free radicals. Other reducing agents will be needed to restore the state, and enzymes which can catalyze the reduction reaction will be involved. Here is another example; glutathione is the most important intracellular reducing agent, which is also the essential substrate catalyzed by glutathione peroxidase reduction, so the concentration of reduced glutathione usually is 10 times more than that of oxidized glutathione. NADPH provides the reducing equivalents for biosynthetic reactions and the oxidation–reduction involved in protecting against the toxicity of ROS, allowing the regeneration of reduced glutathione. The major source of NADPH in animals and other nonphotosynthetic organisms is the pentose phosphate pathway. In order to maintain its antioxidant capacity such as vitamin C, vitamin E, and glutathione, reducing capacity is needed, which comes from energy metabolism. Therefore, the so-called antioxidant is just a part of the cellular antioxidant network.

Since most ROS are bioactivators, when the concentration of ROS exceeds a certain level, it may result in adverse effects on other biological molecules. But most oxidative damage caused by ROS is relatively minor, which is not the key to oxidative damage. These ROS can be converted into highly toxic ROS. For example, hydroxyl radicals (·OH) are highly reactive, nonselective oxidants. When biological systems are exposed to hydroxyl radicals, they can cause damage to cells, including those in humans, where they react with DNA, lipids, and proteins. In fact, this molecule is usually the source of oxidative damage we discussed before. There exist many other ROS such as peroxynitrite, which has the same oxidation damaging capacity.

In conclusion, the oxidative damage is not simply caused by increase of ROS, oxidative stress, or lack of antioxidation capacity. Oxidative stress reflects an imbalance between the systemic manifestation of ROS and a biological system's ability to readily detoxify the reactive intermediates or to repair the resulting damage.

5.3 Redox Balance System

The mainstream view is that there are two types of antioxidant system. One is an antioxidant enzyme system, including superoxide dismutase, catalase, glutathione peroxidase. The other is a nonenzymatic antioxidant system, including vitamin C; vitamin E; glutathione; melatonin; α-lipoic acid; carotenoids; trace elements such as copper, zinc, selenium (Se); and so on. These two systems are not isolated, but can be considered as a whole system. While the antioxidant enzyme system plays its catalytic antioxidant role, the nonenzymatic antioxidant system provides substrates for antioxidation reaction.

Currently, antioxidant theory is based on all free radicals being toxic. Since free radicals are important bioactivators, the name "antioxidant system" may be misleading. I recommend calling it "redox balance system" instead. This is similar to the acid–base balance system; in order to maintain a stable pH balance of the body, the body uses a buffer system, without which wide fluctuations in acid–base balance will appear and may lead to cell dysfunction. Redox is also a balance system, which is maintained by buffering as well.

5.3.1 The Nature of Antioxidation

We have already mentioned that oxygen is the only final electron acceptor in the body, which can also be considered as the ultimate source of in vivo oxidation. How much oxygen do we consume daily? It depends on different active states, for example, oxygen consumption during sleep is 0.24 L/min, and the oxygen consumption during strenuous exercise is 2.8 L/min. Therefore, the daily oxygen consumption ranges from 350 to 4000 L. As much as 2% oxygen is thought to be converted into ROS; according to this ratio, the total daily production of ROS in the body is

7 L out of 80 L oxygen. What scale is this? This is equal to 10 and 114 g, or 0.3 and 3.6 mol in quantity.

Let us have a look at the data on vitamin C first. The normal quality of vitamin C in the body's metabolic pool is about 1.5 g, and its maximum storage can be 3 g, or 0.009 and 0.018 mol. Then let us take a look at the data on vitamin E. Taking relatively high doses of vitamin E (D-alpha-tocopherol 400–800 mg/day) for many years will not do any apparent damage to adults; taking dosage of 800–3200 mg/day will occasionally lead to muscle weakness, fatigue, vomiting, and diarrhea. The most obvious toxic effects after taking dosage more than 1000 mg/day is the antagonistic effect of vitamin K itself, which enhances the anticoagulant effect of orally taking coumarin and results in significant bleeding. Many people believe that vitamins E, C, and A are important exogenous antioxidants; supplementation with these vitamins is an effective means against oxidative damage propagated in many multivitamin products. By the above calculation, it is easy for us to find out even all the vitamins in the body are consumed in the same time (which is impossible, because it will lead to serious disorders), it still have no impact on the total amount of ROS.

There is a question of whether the body has a higher concentration of endogenous antioxidants, such as glutathione. The prevailing view is that glutathione is an important endogenous antioxidant, whose most important role is to maintain various protein thiols in the redox state. In fact, even if all sulfhydryl proteins in vivo are calculated, which includes all the glutathione, the total concentrations will be below 5 mM. In accordance with the maximum concentration, the thiol content of the body is only up to 0.2 mol, which is currently recognized as the maximum antioxidant reserve *in the human body.*

According to the current view of antioxidants, based on the above data, it can be said that ROS produced in the body could not be effectively removed even in the resting state. If so, no one can survive, but in fact no significant oxidative damage is found. Why would the well-recognized antioxidant theory come to this absurd inference?

This relates to the most important issues in this chapter: What is the nature of the antioxidant, or what is the source of the antioxidant *in the human body*? We mentioned before that the real source of antioxidants is the energy substances taken in or stored in the body, including sugar, fat, and protein. The Krebs cycle is the center for energy metabolism which comes after the link reaction and provides the hydrogen and electrons needed for the electron transport chain; these electrons enter into oxidative phosphorylation and produce energy for the cell directly. In this process, the continuous generation of electron transfer is the source of reduction; 89 % of the oxygen was reduced by electrons in oxidative phosphorylation, and 2 % of oxygen converted into ROS, which will also be neutralized by reduction.

When a small quantity of intracellular hydrogen peroxide was produced, glutathione peroxidase will reduce lipid hydroperoxides to their corresponding alcohols and reduce free hydrogen peroxide to water, but glutathione will be oxidized to its oxidized form. It should be particularly noted that, although ROS are neutralized, and the intracellular oxidative stress will not decrease, the capacity is transmitted to the oxidized glutathione. Since oxidized glutathione exists in the liver and erythrocytes, which can be catalyzed by glutathione reductase, it accepts electron from

NADPH, and makes it possible for the ROS elevation reaction to continue. Please note that NADPH offers electrons; where does NADPH get the electrons? NADPH, abbreviated NADP+ or, in older notation, TPN (triphosphopyridine nucleotide), is a cofactor used in anabolic reactions, such as lipid and nucleic acid synthesis, which require NADPH as a reducing agent. The pentose phosphate pathway (also called the phosphogluconate pathway and the hexose monophosphate shunt) is a biochemical pathway parallel to glycolysis that generates NADPH and pentoses (five-carbon sugars). In other words, in order to maintain the oxidation state of glutathione, the cells need to obtain reducing power from energy metabolism, which is the nature of the antioxidant in vivo.

5.3.2 The Regulation of the Antioxidant System

We pay attention to two types of antioxidant system: the enzymatic antioxidant system and the nonenzymatic antioxidant system. The nature of antioxidant systems is not the antioxidant effects, but transfer electron to ROS produced by energy metabolism. In these reactions, enzymatic antioxidant systems promote transmission speed, whereas nonenzymatic antioxidant systems work as substrate for electron transmission. From the perspective of electron transfer, these two antioxidant systems play the same role.

When oxidative stress is relatively low, the antioxidant system can play an effective electron transfer task which ensures the completion of the antioxidant reaction. However, when oxidative stress increases, such as strenuous activity, ROS in vivo will increase a lot; cells need to make use of antioxidant capacity more effectively. So is there such a system for regulating antioxidant capacity?

There really exist systems which can act as a regulator of the total antioxidant system which is called nuclear factor (erythroid-derived 2)-like 2 systems, also known as Nrf2. The Nrf2 antioxidant response pathway is "the primary cellular defense against the cytotoxic effects of oxidative stress." Among other effects, Nrf2 increases the expression of several antioxidant enzyme such as SOD, Cat, HO-1 enzyme as well as endogenous antioxidant glutathione synthetase of the nonenzymatic system. Under oxidative stress, Nrf2 is not degraded, but instead travels to the nucleus where it binds to a DNA promoter and initiates transcription of the above antioxidative genes and their proteins. The existence of Nrf2 indicates the antioxidant defense system can be automatic adjusted as a whole after oxidative stress [29].

The truth is that endogenous antioxidants can be synthesized by ourselves, which indicates the body has its own antioxidant capacity and this capacity is far more than that of "exogenous" antioxidants. In addition, various types of vitamins are exogenous antioxidants for human body, but for many animals and plants they can be synthesized by themselves from energy and energy substances. The ability of synthesis of vitamins has been lost by humans through evolution.

As long as there are no obvious deficiencies of these exogenous antioxidants, actually there is no need to supplement with antioxidants, because supplementation with too many exogenous antioxidants will cause the oxidative stress to decrease, which will inevitably lead to an overall reduction of the antioxidant system and be unfavorable for effective antioxidation. The more reasonable approach is to maintain a certain level of oxidative stress, while utilizing oxidative stress to arouse endogenous antioxidant systems.

5.3.3 The Mobilization of Endogenous Antioxidant Capacity

Many studies have found that various types of vitamin supplements alone cannot effectively play the role of antioxidant, while the use of natural fruits and vegetables can be more effective. What is the real reason? There are some minor toxic substances in the vegetables and fruits we eat, such as broccoli which is rich in sulfur, the most effective agonist of Nrf2 ever found. Sulfur itself is not reductive, but has a weak oxidation which can promote Nrf2 activity and arouse endogenous antioxidant capacity. Many studies have shown that substances such as sulfur has great anticancer and antiinflammatory effect which lasts for a very long time [30]. No such effect has ever been found in vitamins C and E.

People gradually obtain knowledge about the mobilization of endogenous antioxidants with the progress of the study on redox. You can understand the evolution of this knowledge more clearly from the history of free radical theory.

In free radical theory of aging, aging is believed to be accumulated by free radical damage with time. Most free radical damage is oxidative damage in aerobic biological systems. Antioxidants are substances with reduction through which oxidative damage is limited by counteracting free radicals. Strictly speaking, the free radical theory focuses not just on the role of free radicals but also on the role of various ROS such as hydrogen peroxide. In the 1950s, Denham Harman first proposed this theory; this theory was further improved 20 years later in that ROS produced by mitochondria was thought to be the cause of cellular aging. Indeed, there is research evidence showing that reducing oxidative damage can extend the life of yeast and fruit flies, while enhancing oxidative damage can shorten the life of mice. But recently, studies have found that by blocking antioxidant enzyme SOD can prolong the life of *Caenorhabditis elegans* [31]. Therefore, it is still not confirmed that life span can be extended by reducing oxidative damage.

In the 1950s, Dr. Harman proposed the free radical theory, originating from two bases [32]. One basis is the theory of life rate which thought life is inversely proportional to oxygen consumption and metabolic rate. Another basis is that Harman found hyperbaric oxygen toxicity and radiation damage could both be explained by the increase in oxygen free radicals. Taking into account that the radiation causes gene mutations, cancer, and aging, Harman, therefore, concluded free radicals produced during energy metabolism could cause oxidative damage, which leads to organ dysfunction and eventually death with long-term accumulation. Later, the free radical theory was extended from explaining aging to age-related diseases. Intracellular oxidative damage are involved in many types of diseases, such as cancer, arthritis, arteriosclerosis, diabetes, senile dementia, and so on. Many fundamental pathophysiological processes such as phagocytosis, inflammation, and apoptosis are involved in the increase of free radicals. Cell apoptosis is an important means of cell death control, which has a close relationship with the increase of free radical or redox signaling system. Redox-related signals also have an important role in other types of cell death such as necrosis and autophagy. Recently, studies on free radical-related diseases have led to two types of views. In one view, aging is thought to be caused by free radicals directly. In the other, aging-related diseases are caused by free radicals; the accumulation of such diseases is the source of aging.

Here are some experimental evidences of the free radical theory. It was found that overexpression of catalase in mice can extend 20% of maximum life [33]. But the results could not be repeated in a later study in 2009, in which overexpression of catalase and SOD had no effect on the lifespan of mice [34]. Therefore, this experimental evidence is recognized. Other studies showed that the same way would shorten the life expectancy of free-radical-sensitive nematodes while extending the life expectancy of the free-radical-insensitive nematodes [35]. Studies on *Drosophila* also showed that ROS-related metabolism enzyme deficiencies could significantly shorten the life span, while the sensitivity to oxidative damage and radiation damage was significantly enhanced [36].

In recent research on aging, the excitotoxicity hypothesis was applied. According to the excitotoxicity hypothesis, free radicals could promote antioxidant capacity appropriately to reduce oxidative damage, which resist aging. [37] In other words, the latest antiaging theory has changed from antioxidation to mobilization of the endogenous antioxidant system. This latest theory is the summary of the studies of aging research. After long-term studies, it was well established that the only effective means to antiaging is calorie restriction. After calorie restriction, histone deacetylase *in the human body* is activated, which is an important mechanism in extending the life span [38]. Recent studies have found that many substances which can activate histone deacetylase have weak oxidizing capabilities; calorie restriction itself is an external stimuli of oxidative stress. These studies led to the recognition that the use of relatively mild adverse stimulus is the most effective means to extend the life span [39].

In conclusion, simply supplementing with antioxidant supplements is a passive means, which not only cannot really have antioxidant effect but also suppresses the oxidative level which will weaken the stimulating signal of antioxidation, resulting in decline of endogenous antioxidant capacity. The ideal means to deal with oxidative damage is the mobilization of endogenous antioxidant systems by increasing the level of oxidative stress, such as proper physical training and a balanced diet. The intake of substances which can induce the endogenous antioxidant system from natural fruits and vegetables is recommended.

5.4 Members of ROS and Selective Antioxidation

ROS are chemically reactive molecules containing oxygen, which has a strong oxidizing or reducing ability; typical examples include superoxide anion radical, hydroxyl radical, peroxide, nitric oxide, etc., which belong to the family of free radicals; other ROS are common molecules such as hydrogen peroxide, lipid hydroperoxide, hypochlorous acid, etc.

5.4.1 Source of ROS

There are many ways for the generation of ROS in the body; the main ways are listed below:

(1) Xanthine oxidase

Xanthine oxidase (XO, sometimes "XAO") is a form of xanthine oxidoreductase, a type of enzyme that generates ROS. These enzymes catalyze the oxidation of hypoxanthine to xanthine and can further catalyze the oxidation of xanthine to uric acid. These enzymes play an important role in the catabolism of purines in some species, including humans.

(2) Protein kinase C

Protein kinase C (PKC) catalyzes the oxidation of NADPH to NADP + and superoxide anion.

(3) Myeloperoxidase

Myeloperoxidase (MPO) is most abundantly expressed in neutrophil granulocytes. MPO produces hypochlorous acid (HOCl) from hydrogen peroxide (H_2O_2) and chloride anion (Cl^-; or the equivalent from a nonchlorine halide) during the neutrophil's respiratory burst. Furthermore, it oxidizes tyrosine to tyrosyl radical using hydrogen peroxide as an oxidizing agent.

(4) Nitric oxide synthase

NOSs are a family of enzymes catalyzing the production of NO from L-arginine, which is widely distributed in vascular endothelial cells, platelets, neural cells, etc.

The above reactions can be thought of as the source of ROS; when these ROS are generated, they can be converted into other types of ROS by catalytic enzymes or through direct secondary reaction. For example, SOD can catalyze the dismutation of superoxide (O^{2-}) into oxygen and hydrogen peroxide. In the presence of divalent iron ions, hydrogen peroxide can be oxidized to hydroxyl radicals.

In a view, oxygen is the source of ROS production in vivo; it is impossible to produce ROS without oxygen. Since oxygen is the basis for the survival of aerobic organisms, ROS will be inevitably produced at the same time. It can be said that aerobic metabolism must be accompanied by ROS production. When there is no ROS, there will be no oxygen and no aerobic organisms as well.

5.4.2 Types of ROS

There are many types of ROS. Listing all the ROS alone will not be helpful for us to further understand the role of ROS. But appropriate division of ROS will do some help. It can be divided into radical ROS and nonradical ROS according to whether they are free radicals. Radical ROS include superoxide anion radicals, hydroxyl radicals, nitric oxide, etc.; nonradical ROS includes hydrogen peroxide, hypohalous acid, etc. According to whether they contain nitrogen, ROS can be divided into ROS and reactive nitrogen species. Therefore, reactive nitrogen species is one type of ROS.

If ROS are classified by toxicity or activity, they can be divided into three categories. One category is those that can act as signal molecules, e.g., nitric oxide, superoxide anion, and hydrogen peroxide. This type of ROS will be toxic only in very high concentrations, or they are converted into more toxic ROS; they act signal molecules under normal physiological conditions. The second category is those that mainly act as toxic ROS, e.g., hydroxyl radicals, hypochlorous acid, nitrous acid anion, etc. These ROS have strong activity, the concentration of which is low *in the human body*; it will cause damage to the body when they are produced a lot. Many nontoxic ROS become toxic when they are converted into such ROS. For example, the reaction of nitric oxide and superoxide anion produces peroxynitrite anion; superoxide anion and hydrogen peroxide can be the converted into hydroxy radical in the presence of metal ions. The third category is those without signal effects and toxic effects, including the secondary products of the ROS-producing reactions and biological macromolecules. The content of these ROS is relatively low, or is insufficient to produce significant toxic or signal effects. These ROS get relatively less attention; some effect of them may not have been found. According to this classification, we can update our understanding of ROS, or even reevaluate the past antioxidant therapy strategy. It is to say that we should not interfere with the normal physiological functions through overscavenging ROS, especially for normal people; strong antioxidants such as vitamins, pharmaceuticals, or food should not be supplemented. Numerous studies show that no antiaging and antitumor effects have been found after long-term use of various types of antioxidant vitamins; sometimes, it may promote the process of aging and cancer.

5.4.3 The Content of ROS and Its Half-Life

Under normal circumstances, free radicals are constantly produced in the metabolic process. For instance, oxidative phosphorylation is a typical free radical reaction. So free radical reaction is an important basis for life, and in the process of reaction, some of the intermediate product is released to the surroundings which are free radicals and can be converted into ROS.

Different types of ROS *in the human body* are often described by means of oxygen metabolism. The main means of oxygen metabolism in vivo is to promote oxidative phosphorylation. During oxidative phosphorylation, electrons are transferred from electron donors to electron acceptors (one oxygen molecule receives four electrons and combines with two hydrogen ions to form water). Simply speaking, while obtaining four electrons, an oxygen molecule is reduced to water. However, in cells, about 2–3 % oxygen cannot obtain four electrons, but only one electron (not two), which converts oxygen into superoxide anion. Such a reaction widely occurs in mitochondrial oxidative phosphorylation when electrons leak out, e.g., oxygen is directly reduced by coenzyme Q. Superoxide anion is a typical free radical, and an ion as well. Since ions cannot freely pass through the cell membrane, they easily build up high local concentrations, which cause oxidative damage.

In the long-term evolutionary process, all the cell parts where superoxide anion is produced also have the function of generating SOD, which can quickly catalyze superoxide anion into hydrogen peroxide. Hydrogen peroxide is fat soluble and can freely pass through the cell membrane; such an evolutionary change could reduce the oxidative damage caused by local gathering of superoxide anion, which plays a detoxification or antioxidant role. The mechanism is by using the diffusion capacity of hydrogen peroxide to reduce the local concentration. Due to the wide distribution of intracellular SOD and its strong activity, the intracellular concentration of hydrogen peroxide is 1000 times that of superoxide anion. Therefore, from the view of concentration, hydrogen peroxide is truly representative of ROS in the cells. There are enzymes in cells which convert hydrogen peroxide. For example, catalase may convert hydrogen peroxide into water and oxygen. Thus, this is a perfect detoxification process; the superoxide anion can be converted into hydrogen peroxide, and then finally converted into water and oxygen. However, under pathological conditions of hypoxia or inflammation, there will be excessive intracellular free metal ions, such as iron or copper, in the presence of which hydrogen peroxide will become the highly active ROS hydroxyl radical and then oxidative damage will be inevitable. Some cells, such as macrophages, in the process of inflammatory response can not only produce large amounts of superoxide anion but also produce large amounts of nitric oxide by the activation of inducible NOS. The reaction of nitric oxide and superoxide anion will produce another more toxic ROS, nitrite anion.

The first category of ROS has a relatively weak oxidation capacity and is relatively inactive, and its concentration in vivo is often relatively high, such as nitric oxide and hydrogen peroxide whose concentration can reach 10^{-9} M. Superoxide anion is an exception, since a large amount of SOD exists in cells; its concentration is 10^{-12} M. The second category of ROS has powerful toxicity, such as hydroxyl radicals, whose concentration is very low, and cannot be detected in conventional detection methods (Table 5.1).

Table 5.1 Half-life and physiological concentrations of different ROS

ROS	Half-life (s)	Physiological concentrations (M)
Hydroxyl radicals	10^{-9}	Not detected
Nitrite anion	0.05–1.0	Not detected
Nitric oxide	1–10	10^{-9}
Peroxide	Hours or days accelerated by enzymatic	10^{-9}–10^{-7}
Superoxide anion	Hours or days 10^{-6} in the presence of SOD	10^{-12}–10^{-11}

ROS reactive oxygen species, *SOD* superoxide dismutase

According to the above knowledge of free radicals and ROS, we understand they are beneficial to the body. Under the state of ischemia and inflammation, free radicals or ROS increase in concentration, particularly the highly toxic free radicals, and then oxidative damage occurs. This is an important pathophysiological basis of many diseases, and the mobilization of endogenous antioxidant capacity is effective in reducing oxidative damage, which is a potential strategy in the treatment of many diseases. In addition, there are many types of free radicals or ROS, most of which are functional molecules beneficial to the body; only a small part of them is active enough to cause damage to the body.

Currently the academics think selective antioxidation is the main mechanism underlying hydrogen's effect, which is based on the observation that hydrogen can neutralize toxic free radicals such as hydroxyl radicals and nitrite anion while not disturbing those ROS which have biological activity such as nitric oxide hydrogen peroxide and superoxide anions [40] (Fig. 5.1). In recent years, in-depth research on the biological effects of the hydrogen has been carried out. According to the findings, various mechanisms have been proposed, such as the signaling molecule hypothesis that hydrogen gas is the fourth gasotransmitter like nitric oxide [41]. However, this hypothesis is based on the finding that hydrogen can affect some important signaling pathways, but no clear details of how hydrogen affect these pathways has been confirmed. Some scientists have also proposed the gene regulation hypothesis based on the finding that hydrogen could affect the expression of certain genes; the evidence is not enough [42, 43]. In conclusion, although there are doubts and defects, selective antioxidation is still well recognized as the mechanism of hydrogen.

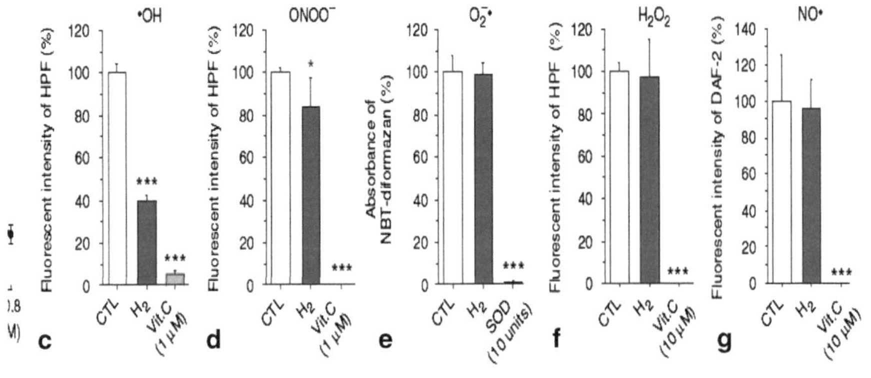

Fig. 5.1 Selective antioxidation of hydrogen. Hydrogen can neutralize hydroxyl radicals and anion nitrite, while not disturbing those ROS which has biological activity such as nitric oxide hydrogen peroxide and superoxide anions [40]

References

1. Halliwell B, Gutteridge J. Free radicals in biology and medicine. New York: Pergamon; 1985.
2. Rao A, HeanáKim K, HanáAhn K. A turn-on two-photon fluorescent probe for ATP and ADP. Chem Commun. 2012;48:3206–8.
3. Balaban RS. Regulation of oxidative phosphorylation in the mammalian cell. Am J Physiol. 1990;258:C377–89.
4. Dröge W. Free radicals in the physiological control of cell function. Physiol Rev. 2002;82:47–95.
5. Furchgott RF, Zawadzki JV. The obligatory role of endothelial cells in the relaxation of arterial smooth muscle by acetylcholine. Nature. 1980;288:373–6.
6. Hutchinson PJ, Palmer RM, Moncada S. Comparative pharmacology of EDRF and nitric oxide on vascular strips. Eur J Pharmacol. 1987;141:445–51.
7. Palmer RM, Ashton D, Moncada S. Vascular endothelial cells synthesize nitric oxide from L-arginine. Nature. 1988;333:664–6.
8. Wood AJ, Parker JD, Parker JO. Nitrate therapy for stable angina pectoris. N Engl J Med. 1998;338:520–31.
9. Garthwaite J. Glutamate, nitric oxide and cell-cell signalling in the nervous system. Trends Neurosci. 1991;14:60–7.
10. Schuman EM, Madison DV. A requirement for the intercellular messenger nitric oxide in long-term potentiation. Science. 1991;254:1503–6.
11. Bult H, Boeckxstaens G, Pelckmans P, Jordaens F, Van Maercke Y, Herman A. Nitric oxide as an inhibitory non-adrenergic non-cholinergic neurotransmitter. Nature. 1990;345:346–7.
12. Smet P, Jonavicius J, Marshall V, De Vente J. Distribution of nitric oxide synthase-immunoreactive nerves and identification of the cellular targets of nitric oxide in guinea-pig and human urinary bladder by cGMP immunohistochemistry. Neuroscience. 1996;71:337–48.
13. Bogdan C. Nitric oxide and the immune response. Nat Immunol. 2001;2:907–16.
14. Fridovich I. Superoxide anion radical ($O.^{-}_{2}$), superoxide dismutases, and related matters. J Biol Chem. 1997;272:18515–7.
15. Veal EA, Day AM, Morgan BA. Hydrogen peroxide sensing and signaling. Mol Cell. 2007;26:1–14.
16. Babior B. The respiratory burst of phagocytes. J Clin Invest. 1984;73:599–601.
17. Suzuki N, Miller G, Morales J, Shulaev V, Torres MA, Mittler R. Respiratory burst oxidases: the engines of ROS signaling. Curr Opin Plant Biol. 2011;14:691–9.
18. Buonocore G, Perrone S, Tataranno ML. Oxygen toxicity: chemistry and biology of reactive oxygen species. Semin Fetal Neonatal Med. 2010;15(4);186–90.
19. Zoungas S, Patel A, Chalmers J, de Galan BE, Li Q, Billot L, Woodward M, Ninomiya T, Neal B, MacMahon S. Severe hypoglycemia and risks of vascular events and death. N Engl J Med. 2010;363:1410–8.
20. Szydlowska K, Tymianski M. Calcium, ischemia and excitotoxicity. Cell Calcium. 2010;47:122–9.
21. West JB. Joseph Priestley, oxygen, and the enlightenment. Am J Physiol Lung Cell Mol Physiol. 2014;306:L111–9.
22. Edwards ML. Hyperbaric oxygen therapy. Part 1: history and principles. J Vet Emerg Crit Care. 2010;20:284–8.
23. Tidwell T. Sunlight and free radicals. Nat Chem. 2013;5:637–9.
24. McCord JM, Fridovich I. Superoxide dismutase an enzymic function for erythrocuprein (hemocuprein). J Biol Chem. 1969;244:6049–55.
25. Babior BM, Kipnes RS, Curnutte JT. Biological defense mechanisms. The production by leukocytes of superoxide, a potential bactericidal agent. J Clin Invest. 1973;52:741–4.
26. Granger D, Höllwarth M, Parks D. Ischemia-reperfusion injury: role of oxygen-derived free radicals. Acta Physiol Scand. Supplementum 1985;548:47–63.
27. Harman D. Aging: a theory based on free radical and radiation chemistry. Berkeley: University of California Radiation Laboratory Berkeley; 1955.

28. Sohal R, Allen R. Oxidative stress as a causal factor in differentiation and aging: a unifying hypothesis. Exp Gerontol. 1990;25:499–522.
29. Kensler TW, Wakabayashi N, Biswal S. Cell survival responses to environmental stresses via the Keap1-Nrf2-ARE pathway. Annu Rev Pharmacol Toxicol. 2007;47:89–116.
30. Vasanthi HR, Mukherjee S, Das DK. Potential health benefits of broccoli-a chemico-biological overview. Mini Rev Med Chem. 2009;9:749–59.
31. Honda Y, Honda S. The daf-2 gene network for longevity regulates oxidative stress resistance and Mn-superoxide dismutase gene expression in *Caenorhabditis elegans*. FASEB J. 1999;13:1385–93.
32. Harman D. Free radical theory of aging: an update. Ann N Y Acad Sci. 2006;1067:10–21.
33. Schriner SE, Linford NJ, Martin GM, Treuting P, Ogburn CE, Emond M, Coskun PE, Ladiges W, Wolf N, Van Remmen H. Extension of murine life span by overexpression of catalase targeted to mitochondria. Science. 2005;308:1909–11.
34. Pérez VI, Van Remmen H, Bokov A, Epstein CJ, Vijg J, Richardson A. The overexpression of major antioxidant enzymes does not extend the lifespan of mice. Ageing Cell. 2009;8:73–5.
35. Cao Z, Lindsay JG, Isaacs NW. Mitochondrial peroxiredoxins. In: Flohé L, Harris JR, editors. Peroxiredoxin systems. Berlin: Springer; 2007. pp. 295–315.
36. Parkes TL, Elia AJ, Dickinson D, Hilliker AJ, Phillips JP, Boulianne GL. Extension of *Drosophila* lifespan by overexpression of human SOD1 in motorneurons. Nat Genet. 1998;19:171–4.
37. Schulz JB, Henshaw DR, Siwek D, Jenkins BG, Ferrante RJ, Cipolloni PB, Kowall NW, Rosen BR, Beal MF. Involvement of free radicals in excitotoxicity in vivo. J Neurochem. 1995;64;2239–47.
38. Chang KT, Min K-T. Regulation of lifespan by histone deacetylase. Ageing Res Rev. 2002;1:313–26.
39. Bordone L, Guarente L. Calorie restriction, SIRT1 and metabolism: understanding longevity. Nat Rev Mol Cell Biol. 2005;6:298–305.
40. Ohsawa I, Ishikawa M, Takahashi K, Watanabe M, Nishimaki K, Yamagata K, Katsura K, Katayama Y, Asoh S, Ohta S. Hydrogen acts as a therapeutic antioxidant by selectively reducing cytotoxic oxygen radicals. Nat Med. 2007;13:688–94.
41. George JF, Agarwal A. Hydrogen: another gas with therapeutic potential. Kidney Int. 2010;77:85–7.
42. Chen H-G, Xie K-L, Han H-Z, Wang W-N, Liu D-Q, Wang G-L, Yu Y-H. Heme oxygenase-1 mediates the anti-inflammatory effect of molecular hydrogen in LPS-stimulated RAW 264.7 macrophages. Int J Surg. 2013;11:1060–6.
43. Iio A, Ito M, Itoh T, Terazawa R, Fujita Y, Nozawa Y, Ohsawa I, Ohno K, Ito M. Molecular hydrogen attenuates fatty acid uptake and lipid accumulation through downregulating CD36 expression in HepG2 cells. Med Gas Res. 2013;3:6.

Chapter 6
Therapeutic Effects of Hydrogen on Different Diseases

Liren Qian, Jianliang Shen and Xuejun Sun

Abstract Since 2007, hydrogen gas biology has become hotspot in medical research because of its selective antioxidant effects reported by Ohsawa et al. By the end of 2014, more than 400 papers have been published in related area, including organ ischemia-reperfusion injury, diabetes, atherosclerosis, hypertension, cancer, and other major human diseases. Many of these studies were based on the premise that the hydrogen gas is a new type of antioxidant substance, and also there are a lot of evidences of its antiapoptotic, anti-inflammatory effects. Since oxidative stress is related to almost all cells, tissues, and organs, it is related to almost all diseases. In this chapter, we focus on some typical types of diseases, including nervous system disorders, liver diseases, metabolic diseases, and some clinical studies in order to facilitate the readers to understand the molecular biology of hydrogen.

Keywords Hydrogen · Therapy · Nervous system · Metabolic · Liver · Ophthalmic · Respiratory · Inflammatory · Translational

In 2007, Ohsawa et al. found the selective antioxidative and therapeutic effects on cerebral ischemia of hydrogen gas [1]. From 2007 to now, hydrogen gas was demonstrated effective in more than 60 kinds of human diseases, most of which are oxidative stress-related diseases, including neonatal cerebral hypoxia, Parkinson's disease, and tissue ischemia-reperfusion (I/R; spinal cord, heart, lung, liver, kidney, and small intestine) [2]. Despite there are many doubts, selective antioxidative effect remains the main mechanism of hydrogen which is currently accepted. In addition, its antiapoptotic, anti-inflammatory effects play important roles in studying its biological effects.

Although there were reports of antioxidative effects of hydrogen gas as early as 1975 and 2001 [3, 4], the hydrogen gas used in the study in 2001 was with 800 kPa and 14 days, while the hydrogen gas used in the study in 2007 was with 2 kPa and

X. Sun (✉)
Department of Navy Aeromedicine, Second Military Medical University, Shanghai, China
e-mail: sunxjk@hotmail.com

L. Qian · J. Shen
Department of Hematology, Navy General Hospital, Fucheng Road, Beijing, China

X. Sun et al. (eds.), *Hydrogen Molecular Biology and Medicine*,
DOI 10.1007/978-94-017-9691-0_6

less than 1 h [1]. The two studies were definitely different with entirely different nature. The article in 2007 attracted international attention rapidly after its publication. The number of articles about hydrogen biology was with explosive growth after its publication. The paper in 2007 has been cited more than 300 times. Although most of these papers were trying to prove the therapeutic effects of hydrogen, there were two papers that reported its ineffectiveness on severe neonatal hypoxia–ischemia and disuse muscle atrophy [5, 6].

Conventional treatment options are always ineffective for severe diseases. Hydrogen is also not an exception. Although it was suggested that oxidative stress is involved in the pathogenesis of disuse muscle atrophy [7], hydrogen was ineffective on this kind of disease, which suggest that oxidative stress may not be the most important pathogenesis of disuse muscle atrophy. Therapeutic effects of hydrogen on organ I/R injury and inflammation-related diseases are the most significant.

Hydrogen is the simplest atom in nature, which consists of a proton and extranuclear electron. Hydrogen gas is the simplest molecule in nature that is composed of two hydrogen atoms. When hydrogen gas and oxygen gas are mixed, combustion or explosion may occur at concentration ranging from 4 to 75 %. However, hydrogen gas is chemically inactive at room temperature and cannot react with other substances easily. Despite of this, hydrogen gas is a small molecule that can easily diffuse into body and cells, leading to high probability of collision between hydrogen and other atoms or molecules. The powerful diffusion capability of hydrogen gas perhaps makes up for the low reaction rate of hydrogen gas. This makes hydrogen gas not only stable but also selectively antioxidant, which is the special advantage of hydrogen gas.

Reactive oxygen species (ROS) play an important role in the pathophysiology of various types of cardiovascular and cerebrovascular diseases, such as stroke and myocardial infarction, metabolic diseases such as diabetes, atherosclerosis, and other important human acute and chronic diseases. It exists as two forms, oxygen-free radicals and non-oxygen-free radicals. Oxygen-free radicals include hydroxyl radical, superoxide anion, nitric oxide, nitrite anion, and other substances. Under physiological conditions, ROS are continuously generated in the body and are also continuously cleared, which is in dynamic equilibrium. However, under pathological conditions such as ischemia and inflammation, a large amount of ROS may be produced. Hydroxyl radicals and peroxynitrite are the main substances that make oxidative damage to cells. The toxicities of nitric oxide, superoxide anion, and hydrogen peroxide and other substances are weak, but they play an important role in signal transduction. In the treatment of oxidative damage previously, reductive drugs may lead to new imbalance of oxidative-redox state which is the main reason for current poor effects of antioxidant treatments. In 2007, Ohsawa et al. demonstrated that hydrogen could selectively reduce toxic hydroxyl radicals and peroxynitrite anion [1], with little effect on the ROS which have important biological functions with low toxicity. This is the selective antioxidative effect of hydrogen gas which provides a new idea for the antioxidant therapy.

As early as in 2001, Gharib et al. reported inhalation of 8 atm of H_2 has therapeutic effects on liver inflammation caused by *Schistosoma japonicum* infection [4]. They demonstrated that direct reaction between hydrogen and hydroxyl radical is the basis of anti-inflammatory effects of hydrogen. In 2009, Kajiya et al. reported hydrogen could obviously inhibit colonic inflammation induced by dextran sulfate sodium [8]. The anti-inflammatory effect of hydrogen is related to the inhibition of the generation of ROS and proinflammatory cytokine release. In addition, macrophages play an important role in inflammation and immune regulation. Regulation on macrophages laid the foundation for its anti-inflammatory effect.

In 2008, it is found that hydrogen can reduce tissue injury in rats with hypoxia–ischemia model. Inhalation of low concentrations of hydrogen can reduce the activity of caspase-3 and caspase-12 enzyme, which suggested that it has antiapoptotic effect [9]. Because of its antiapoptotic effect, the studies of I/R injury, inflammation, and other diseases are attracting more and more attention. But the exact mechanism of antiapoptotic effect of hydrogen is not completely clear, which needs further study.

ROS play an important role in the angiogenesis process. Studies have shown that ROS can induce vascular smooth muscle hyperplasia. Kubota et al. reported that hydrogen-rich water eye drops have anticorneal angiogenesis effects [10]. The mechanism may be related to its ability of scavenging ROS generated after corneal injury, reduction of nuclear transcription factor gene transcription (NF-κB), and reduction of vascular endothelial growth factor expression levels. Angiogenesis is not only important to the pathophysiology of many vascular diseases but also play important role in the occurrence of tumor. The role of hydrogen on angiogenesis is worthy of further discussion.

6.1 Therapeutic Effects of Hydrogen on Central Nervous System Diseases

Central nervous system is one of the body's most important systems. Central nervous system disorders include acute and chronic central nervous system disorders. Acute central nervous system disorders include trauma and cerebrovascular diseases. Chronic central nervous system disorders include various types of neurodegenerative diseases. Since the biological effects of hydrogen were found, it was demonstrated has significant protective effects on central nervous system dysfunction including cerebral vascular disease, Parkinson's disease, and Alzheimer's disease [11–17]. Preliminary clinical studies have been carried out in cerebral ischemia and Parkinson's disease, which indicate that clinical application of hydrogen in neurological disease attracted great attention [13, 17].

6.1.1 Therapeutic Effects of Hydrogen on Cerebral Vascular Disease

Cerebrovascular disease is a kind of disease involving brain artery or carotid artery, leading to intracranial blood circulation disorders and brain tissue damage. Cerebrovascular disease can be divided into ischemic cerebrovascular disease and hemorrhagic cerebrovascular disease, including cerebral infarction and cerebral embolism. No matter cerebral ischemia or cerebral hemorrhage, there are still a few effective treatments. Although the pathophysiology of cerebrovascular disease is complex, there is a similar pathophysiological process in most types of acute brain injury, such as hypoxia–ischemia, reperfusion injury, calcium overload, free radicals, and inflammatory injury, and so on. Free radicals and inflammation damage are in the heart of the pathophysiology process described above, which is the main reason why hydrogen was studied in various types of cerebrovascular disease. In 2007, Ohsawa et al. reported that inhalation of hydrogen gas exerts therapeutic effects in a rat model with left middle cerebral artery occlusion [1]. After that report, Cai et al. demonstrated that inhalation of hydrogen gas has desired therapeutic effects on hypoxic–ischemic brain damage caused by asphyxia [9].

Although some scholars reported that the therapeutic effects of hydrogen on severe brain hypoxic–ischemic injury in not ideal [5], more studies have shown that hydrogen has protective effects on brain injury caused by cardiac arrest which further confirmed the protective effects of hydrogen on hypoxic–ischemic brain injury [18].

Edaravone is currently the only approved drug for stroke [19]. Ono et al. carried out a clinical research and studied the therapeutic effects of combined hydrogen saturated saline and edaravone on brainstem infarction patients [13], compared with those patients treated with edaravone alone. It was found that magnetic resonance imaging (MRI) indicators of those patients treated with hydrogen and edaravone were better than those treated with edaravone alone. The main mechanism of edaravone is antioxidation, which is very similar to the hydrogen gas. The results suggested, hydrogen has desired therapeutic effects. This is the only clinical study currently reported treating cerebral ischemia by hydrogen.

Intracranial hemorrhage is a kind of cerebrovascular disease which very easily leads to the death, including cerebral hemorrhage and subarachnoid hemorrhage. It has been demonstrated that hydrogen has ideal protective effects on early brain injury, nerve cell necrosis, edema, and vascular spasm caused by cerebral hemorrhage and subarachnoid hemorrhage [20, 21]. This shows that hydrogen has a potential therapeutic value in all kinds of cerebral vascular diseases. Recently, hydrogen was found has therapeutic effects on platelet aggregation which needs to be paid more attention [22], because delayed hemorrhage is a serious complication of cerebral vascular disease which is lethal. But surprisingly, some researchers have proved that cerebral ischemia-induced hemorrhage could be reduced by hydrogen [23]. Apparently, these contradictory issues need further study.

6.1.2 Therapeutic Effects of Hydrogen on Neurodegenerative Diseases

Dopaminergic neuronal death at the substantia nigra pars compact may cause Parkinson's disease, which is always secondary to other neurodegenerative diseases such as Alzheimer's and other diseases. There are two mechanisms that underlie

Parkinson's disease: excessive oxidative stress and abnormal ubiquitin-proteasome system [24].

Dopamine is a kind of neurotransmitter, which is a pro-oxidant by itself, and dopaminergic cells are destined to be exposed to high concentrations of radical oxygen species. Hemi-Parkinson's disease model can be made by stereotactically injecting catecholaminergic neurotoxin 6-hydroxydopamine (6-OHDA) in the right striatum [16]. The development of hemi-Parkinson's symptoms can be completely abolished by administration of hydrogen-rich water 1 week before surgery, and 40.2 % of dopaminergic neurons on the toxin-injected side were reduced, whereas hydrogen treatment improved the reduction to 83.0 %. Hemi-Parkinson's symptoms were also suppressed by giving hydrogen-rich water 3 days after surgery, but not as much as those observed in pretreated rats. Pretreated rats were also sacrificed 48 h after toxin injection and the tyrosine hydroxylase activity at the striatum, where dopaminergic neurons terminate. This indicated that hydrogen did not directly detoxicate 6-OHDA but exerted a delayed protective effect for dopaminergic cells. Fujita et al. also demonstrated a similar prominent effect of hydrogen-rich water on an 1-methyl-4-phenyl-1,2,3,6-tetrahydropyridine (MPTP)-induced mouse model of Parkinson's disease [15]. MPTP is a neurotoxin that blocks complex I of the mitochondrial electron transport system and causes Parkinson's disease in mice and humans. It is interesting to note that the concentration of hydrogen that they used for the MPTP mice was only 0.08 ppm, which is the second lowest among all the trials published to date for rodents and humans. The lowest hydrogen concentration ever tested is 0.048 ppm in the dialysis solution for patients receiving hemodialysis [25].

Alzheimer's disease is the most common neurodegenerative disease which is characterized by abnormal aggregation of β-amyloid (Aβ) and tau, the large aggregates of which are respectively recognizable as senile plaques and neurofibrillary tangles [26]. Three models have been used to study therapeutic effects of hydrogen on Alzheimer's disease. A mouse model of dementia by restricting movement of mice for 10 h a day was made by Nagata et al. [27]. Cognitive functions were analyzed by passive avoidance learning, object recognition tasks, and the Morris water maze. Administration of hydrogen-rich water was demonstrated efficiently ameliorated cognitive impairment. They also showed that neural proliferation in the dentate gyrus was restored by hydrogen. Another rat model was made by intracerebroventricular injection of Aβ1-42 by Li et al. [11]. They found that intraperitoneal injection of hydrogen-rich saline for 14 days efficiently ameliorated cognitive decline and preserved long-term potentiation (LTP) by suppression of abnormal activation of IL1β, c-Jun N-terminal kinases (JNK), and NF-κB [12]. The third model was established by using a senescence-accelerated mouse prone 8 (SAMP8) strain which can exhibit early aging syndromes, including impairment in learning ability and memory [28]. The administration of hydrogen-rich water was demonstrated efficiently prevented cognitive decline by the Morris water maze.

Chorea is a kind of relatively common neurodegenerative disease. Its clinical manifestation is different from senile dementia and Parkinson's disease, but the pathophysiological processes are similar. There is still no report about the therapeutic effects of hydrogen on this disease.

6.2 Therapeutic Effects of Hydrogen on Metabolic Diseases

6.2.1 *Diabetes*

In 2008, Kajiyama et al. performed a randomized, double-blind, placebo-controlled study in 30 patients with diabetes mellitus type II and 6 patients with impaired glucose tolerance [29]. The patients consumed either hydrogen-rich water or placebo water for 8 weeks, with a 12-week washout period. They measured 13 biomarkers to estimate lipid and glucose metabolisms.

All the biomarkers were favorably changed with hydrogen, but statistical significance was observed only in improvement of electronegative charge-modified low-density lipoprotein (LDL) cholesterol, small dense LDL, and urinary 8-isoprostanes. In four of six patients with impaired glucose tolerance, hydrogen normalized the oral glucose tolerance test. Lack of statistical significance in their studies was likely due to the small number of patients and the short observation period. The lack of statistical significance, however, may also suggest a less prominent effect in human diabetes mellitus compared to rodent models [30, 31].

6.2.2 *Metabolic Syndrome*

An open-label trial in 20 subjects with potential metabolic syndrome was performed by Nakao et al. [32]. Hydrogen-rich water was produced by placing a metallic magnesium stick in water, which yielded 0.55–0.65 mM hydrogen water (70–80 % saturation). The participants consumed 1.5–2.0 L of hydrogen water per day for 8 weeks and showed a 39 % increase in urinary superoxide dismutase [29]; a 43 % decrease in urinary thiobarbituric acid reactive substances (TBARS), a marker of lipid peroxidation; an 8 % increase in high-density lipoprotein (HDL) cholesterol; a 13 % decrease in total cholesterol/HDL cholesterol; and a 13 % decrease in total cholesterol/HDL cholesterol. The aspartate aminotransferase [33] and alanine transaminase (ALT) levels remained unchanged, whereas the gamma-glutamyl transferase (GGT) level was increased by 24 % but was still within a normal range. Although the study was not double blinded and placebo controlled, improvements in biomarkers were much more than those in other hydrogen studies in humans. As this study used a large amount of hydrogen water, the amount of hydrogen might have been a critical determinant. Alternatively, excessive hydration might have prevented the participants from excessive food intake.

6.3 Therapeutic Effects of Hydrogen on Liver Diseases

Early in 2001, there was a report about the protective effects of high-pressure hydrogen gas on liver injury [4]. In 2007, therapeutic effects of hydrogen on hepatic ischemia were further studied [1]. Therefore, the study of hydrogen gas on liver disease is relatively early; the research methods and the types of liver diseases are various.

6.3.1 Early Studies of Hydrogen on Liver Diseases

Hydrogen was applied in the field of liver diseases prominently. Gharib et al. examined the effects of molecular hydrogen on a mouse model of schistosomiasis-associated chronic liver inflammation. The mice breathed hydrogen–oxygen mixed gas for 14 consecutive days (87.5 % hydrogen gas) [28]. Liver function, liver tissue oxidative damage, fibrosis, and inflammatory indexes were observed. They demonstrated that continuous high-pressure hydrogen breath has significant protective effects on liver tissue damage, inflammation, and fibrosis. This study was groundbreaking; it is the first proof that hydrogen has anti-inflammatory, anti-hepatic fibrosis effects. However, there are still two obvious deficiencies in the study. First, the dosage is too large. Second, this study cannot replace academic status of the study which has been published articles in 2007 in *Nature Med* [1]. Because the latter proved breathing 2 % hydrogen for 35 min has biological effects which is much more convenient to use. There is also no reasonable explanation in the study in 2001, which led to little attraction from other researchers.

6.3.2 Therapeutic Effects of Hydrogen on Liver Diseases

Japanese researchers published the first paper that demonstrated the therapeutic effects of low-dose hydrogen on cerebral I/R injury. Subsequently, an article reported the therapeutic effects of hydrogen on liver I/R injury. Fukuda et al. established a rat model of hepatic I/R in 2007 [34]. They found that hydrogen has very significant therapeutic effects on liver ischemic injury by hematoxylin and eosin (HE) staining, malondialdehyde (MDA), and enzyme test. In 2009, Kajiya et al. found that hydrogen has preventive effects on concanavalin-induced hepatitis [8]. In the same year, Tsai et al. found that drinking hydrogen-rich water can protect mice from carbon tetrachloride-induced liver damage [35]. Sun HY et al. demonstrated that hydrogen was effective not only on acute liver damage but also on cirrhosis in three kinds of liver injury model by using GalN/LPS, CCl4, and DEN [36]. Liu Q et al. found that intraperitoneal injection of hydrogen-rich saline could improve the antioxidant capacity of liver and inhibit hepatic inflammatory response, which has great clinical significance [37].

It has been reported that electrolytic water has protective effects on alcoholic liver injury which suggests that hydrogen has a sobering effect. Recent studies demonstrated that drinking hydrogen water for a long time can improve diet-induced fatty liver, not only improving liver function and liver morphology such as fibrosis, but also blocking cellular signaling pathways [38]. When fatty liver becomes a new lifestyle disease, the value of the hydrogen will be obvious. In this study, the authors also demonstrated that long-term consumption of hydrogen water can not only improve fatty liver but also reduce the possibility of the occurrence of liver cancer significantly, which has a greater significance. It is the most direct evidence of hydrogen in the prevention of tumorigenesis.

Another research studied the therapeutic effects of hydrogen on obesity which is related to fatty liver [31]. The study is mainly focused on the hepatic metabolism of fat and proved that hydrogen could promote an important signaling molecule FGF21, which leads to weight loss and exerts therapeutic effects on fatty liver. There are more than 22 members in the family of fibroblast growth factor (FGF) that have a variety of biological activities. FGF21 is a new member of FGF family in the year 2000, which specifically expressed primarily in the liver, kidney, and adipose tissue. Studies have shown that FGF21 is a kind of very meaningful cytokines protecting glucose metabolism homeostasis. Its biological function is mainly reflected in the regulation of glucose and lipid metabolism which has a similar effect as insulin. FGF21 activator is known as a targeting molecule of the next-generation drug treating diabetes. Preliminary clinical studies have been reported that hydrogen has therapeutic effects on diabetes [31].

Many hydrogen product manufacturers have described the therapeutic effects of hydrogen on fatty liver which cannot be objective evidence, but also suggests its effects on the disease. The next step is focused on clinical observation of hydrogen on the disease.

There is only a few clinical research of hydrogen on liver disease that has been reported. Kang et al. performed a randomized placebo-controlled study of hydrogen water in 49 patients receiving radiation therapy for malignant liver tumors. Hydrogen improved the quality of life (QOL) scores. In particular, hydrogen efficiently prevented loss of appetite [39].

The studies above show that, as a selective antioxidant, hydrogen has significant therapeutic effects on hepatic ischemia, hepatitis, fatty liver, and other types of liver diseases, which indicate that hydrogen has very broad prospects on liver diseases.

6.3.3 The Outlook of Hydrogen in the Liver Diseases

Although the study of hydrogen in the liver disease has a certain size, there are still many blind spots. (1) Organ transplantation. Study of hydrogen in the liver transplantation is still a blank.

There are related studies of hydrogen on organ transplantation, including kidney, heart, small intestine, and lung. Although liver transplantation has been widely used in clinic, but there is no study about hydrogen on liver transplantation. By learning from other organ transplant injury research model, protective effects of hydrogen on liver transplantation are worth exploring. (2) Viral hepatitis. There is no report of hydrogen on various types of viral hepatitis research. Although it is hard to say hydrogen has direct antiviral effects, but virus-induced liver inflammation immune therapy can be an important basis for using hydrogen. (3) Clinical research. This is of course important and also a common problem existing in the field of hydrogen biology. Biological effects of hydrogen are lacking clinical research evidence, especially those kinds of autoimmune disease, dermatitis, fatty liver, and obesity; chemical radiation therapy are very worthy to carry out clinical research. Nonalcoholic steatohepatitis and accompanying hepatocarcinogenesis (NASH) is one of the important liver diseases to overcome.

6.4 Therapeutic Effects of Hydrogen on Ophthalmic Diseases and Ear Diseases

6.4.1 Retina Ischemia-Reperfusion Injury

Hydrogen eye drops have been proved effective on retina I/R injury. The study showed potential and broad application prospects of hydrogen eye drops [40]. They prepared H_2-loaded eye drops (0.8 mM, pH 7.2) by dissolving H_2 gas into saline to saturated level and then administered the H_2-loaded eye drops to the ocular surface continuously (4 L/min) during the ischemia or reperfusion periods. The H_2-loaded eye drops could effectively protect the retina from I/R injury by scavenging hydroxyl radicals and have an enormous impact via the topical application of H_2 solution.

6.4.2 Erythematous Skin Diseases

Ono et al. treated four patients of acute erythematous skin diseases with fever and/or pain by H_2-enriched intravenous fluid [41]. All of the four patients received intravenous administration of 500 mL of H_2-enriched fluid in 30 min for more than 3 days except in one patient for only once. Erythema of these four patients and associated symptoms improved significantly after the H_2 treatment and did not recur.

6.4.3 Skin Tumors

Until 1975, Dole et al. [3] reported that hyperbaric hydrogen may be a possible treatment for cancer. They found a marked regression of the skin squamous cell carcinoma by inhalation of a mixture (2.5 % O_2 and 97.5 % H_2) at a total of 0.8 MPa for 2 weeks in a mouse model.

6.4.4 Radiation-Induced Skin Injury

Skin is a biological defense barrier of human body. Radiation injures skin directly by radiation energy or indirectly by free radicals, causing radiodermatitis which occurred in nearly 95 % of patients receiving radiation therapy [42]. There are generally two types of radiodermatitis: acute radiodermatitis (usually occur within 90 days) and chronic radiodermatitis (may occurred over a prolonged period) which is often exhibited by the onset of erythema, swelling, blisters, and ulceration, followed by development of chronic inflammation, necrosis, fibrosis, and lymphedema [43, 44]. In 2012, Guo et al. [45] reported first that hydrogen-rich saline protected against ultraviolet B (UVB) radiation injury, possibly by reducing inflammation and oxidative stress. They demonstrated that hydrogen-rich saline had protective effects by altering the levels of markers, including necrosis factor alpha, IL-1â, IL-6, tissue superoxide dismutase, MDA, and nitric oxide activity, and relieved morphological skin injury against UVB radiation injury on C57BL/6 rats. In 2013, Mei et al. [42] reported the radioprotective effects of hydrogen on skin in vitro and vivo. In the study, our group found hydrogen significantly reduced the severity of dermatitis caused by radiation, accelerated tissue recovery, and reduced the extent of radiation-induced weight loss in rats after a single dose of 15 or 20 Gy radiation. We also found hydrogen protected rats from cumulative doses of 30 Gy delivered in three fractions. In the study, hydrogen also protected immortalized human keratinocytes (HaCaT cells) from radiation-induced injury. In 2013, Rosa Mistica Ignacio et al. [46] demonstrated the protective effect of hydrogen-reduced water (HRW) on UVB-mediated skin injury in hairless mice by balneotherapy. In their study, bathing with HRW significantly reduced the levels of skin damage by decreasing the level of inflammatory cytokines, including IL-1β, IL-6, TNF-α and IFN-γ, as well as increased activity of glutathione peroxidase.

In their study, HRW bathing also protect UV-induced corneocytes damage and ultrastructural changes.

Interestingly, in a recent study by Mi Hee Shin et al. [47], they observed that $H(H_2O)$ (atomic hydrogen surrounded by water molecules) application to human skin prevented UV-induced erythema and DNA damage. In their study, $H(H_2O)$ significantly prevented UV-induced MMP-1, COX-2, IL-6, and IL-1β mRNA expressions in human skin in vivo. They also found that $H(H_2O)$ prevented UV-induced ROS generation and inhibited UV-induced MMP-1, COX-2, and IL-6 expressions, and UV-induced JNK and c-Jun phosphorylation in HaCaT cells.

6.5 Therapeutic Effects of Hydrogen on Respiratory Diseases

Inhaled therapeutic gas is a reasonable approach for the treatment of lung injury as imbalances between ROS and the antioxidant defense system are involved in certain pulmonary pathologic conditions, such as lung inflammation, ventilator-induced lung injury (VILI), and acute respiratory distress syndrome [48–51]. In 2010, Huang et al. showed that inhalation of 2 % hydrogen gas attenuates VILI in a mouse model. Ventilation with 2 % hydrogen in balanced air significantly ameliorated VILI-induced lung injury. This data strongly suggests that inhaled hydrogen gas could therapeutically mitigate VILI via its antioxidant and anti-inflammatory effects [52].

Exposure to high concentrations of oxygen for prolonged periods causes hyperoxic lung injury that leads to respiratory failure. High concentrations of oxygen markedly impair lung function and increase inflammatory responses. Hyperoxia in the presence of hydrogen significantly reduced hyperoxia-induced oxidative stress injury in premature rat type II alveolar epithelial cells, improved the cellular antioxidant capacity, stabilized the mitochondrial membrane potential, and reduced the inhibitory effect of hyperoxia on cell proliferation [53]. Acute lung injury secondary to remote organ damage and followed by deleterious systemic inflammation is a critical event. Mao et al. reported that hydrogen-rich saline treatment attenuated lung injury induced by intestinal I/R injury in a rat model [54]. The lung is one of the organs most susceptible to radiation injury [55]. Radiation pneumonitis is an inflammation of the lungs that occurs when lung or whole body was irradiated. Development of interstitial pneumonitis increases according to radiation dose, especially single-fraction total body irradiation at higher dose rates [56] and higher total lung doses [57, 58]. In 2011, our group hypothesized that hydrogen may be a possible prevention strategy for radiation pneumonitis [59]. In 2011, Terasaki et al. [60] showed that H_2 reduced irradiation-induced OH levels in human lung epithelial cell line A549 cells. They demonstrated that pretreatment of H_2 could reduce the fluorescence intensity of hydroxyphenyl fluorescein in irradiated A549 cells. They demonstrated pretreatment of H_2 reduced the products of oxidative stress, including 4-hydroxy-2-nonenal, 8-hydroxydeoxyguanosine, etc. H_2 could also significantly reduce levels of apoptosis-associated proteins, including Bax and active caspase 3 in irradiated A549 cells after a 24-h incubation with H_2-rich solution. In vivo, they demonstrated that H_2 treatment reduced oxidative stress and apoptosis, measures of acute damage in the lungs of mice within 1 week after whole thorax irradiation by immunohistochemistry and immunoblotting. In their study, H_2 treatment reduced lung fibrosis by chest computed tomography, Ashcroft scores, and type III collagen deposition at 5 months after irradiation.

6.6 Therapeutic Effects of Hydrogen on Inflammatory Diseases

6.6.1 Rheumatoid Arthritis

Rheumatoid arthritis (RA) is a chronic inflammatory disease that affects approximately 1 % of the population. It is characterized by irreversible joint disorder accompanied by destruction of bone and cartilage, which causes serious morbidity. ROS play a central role both upstream and downstream of NF-κB and TNF-α pathways, which are located at the center of the inflammatory response. Among the ROS, the hydroxyl radical is the most harmful, and molecular hydrogen (H_2) is a selective scavenger. It was demonstrated that consumption of water with a high concentration of molecular hydrogen significantly improves the disease activity and reduces the oxidative stress in RA for this species [61, 62]. It has been shown that H_2 is useful when administered along with the conventional therapy in RA as it acts to reduce oxidative stress in the patients. Especially in the early stage, H_2 showed significant therapeutic potential, which also seemed to assist diagnosis and treatment decisions of RA.

6.6.2 *Ulcerative Colitis*

Ulcerative colitis (UC) is an inflammatory bowel disease (IBD) that features a chronic, relapsing, and remitting inflammatory condition of the intestine, which is associated with enhanced production of ROS and altered angiogenesis. It has been reported that the treatment with hydrogen-rich saline reduced the weight loss and diarrhea and alleviated the colonic mucosal damage in the UC rats. In addition, the expression of vascular endothelial growth factor in the UC rats increased and could be inhibited by hydrogen treatment [63].

6.7 Studies of Hydrogen in Translational Medicine

Up to now , there have been seven kinds of diseases reported in clinical studies, including type II diabetes, metabolic syndrome, RA, hemodialysis, inflammation/mitochondrial muscle disease, brainstem ischemia, radiation therapy. In the World Health Organization Registration information, we found there are some clinical studies without published papers. By reading the following description of these studies, clinical researchers can learn more about the research progress of clinical studies of hydrogen.

Biological effects of hydrogen are mainly obtained in rodent models. So far, only seven preliminary clinical studies of hydrogen have been reported. These clinical studies are only preliminary clinical observation and insufficient to support hydrogen as a clinical treatment. Here we reviewed these studies. A hydrogen clinical trial in Parkinson's disease has been carried out which showed good therapeutic effects which has not been published (in fact, Japanese has a large clinical data about cognitive impairment and elderly diabetes). Type II diabetes, metabolic syndrome, and RA have been described above. Here, we describe the other four diseases.

6.7.1 *Hemodialysis*

An open-label placebo-controlled crossover trial of 12 sessions of hemodialysis in 8 patients and an open-label trial of 78 sessions of hemodialysis in 21 patients were performed by Nakayama et al. [25, 64]. In both studies, continuous sessions of hemodialysis with hydrogen-rich dialysis solution decreased systolic blood pressure before and after dialysis. In the short-term study, plasma methylguanidine was significantly decreased. In the long-term study, plasma monocyte chemoattractant protein 1 and myeloperoxidase were significantly decreased.

6.7.2 Muscular Diseases

Ohno et al. performed an open-label trial of hydrogen water in 14 patients with muscular diseases, including muscular dystrophies, polymyositis/dermatomyositis, and mitochondrial myopathies, and a randomized, double-blind, placebo-controlled trial of hydrogen water or dehydrogenized water in 22 patients with dermatomyositis and mitochondrial myopathies [65]. In the open-label trial, significant improvements were observed in lactate-to-pyruvate ratio, fasting blood glucose, serum MMP-3, and triglycerides. Especially, the lactate-to-pyruvate ratio, which is a sensitive biomarker for the compromised mitochondrial electron transport system, was decreased by 28% in mitochondrial myopathies. In addition, MMP-3, which represents the activity of inflammation, was decreased by 27% in dermatomyositis. In the double-blind trial, a statistically significant improvement was observed only in serum lactate in mitochondrial myopathies, but lactate-to-pyruvate ratio in mitochondrial myopathies and MMP-3 in dermatomyositis were also decreased. Lack of statistical significance with the double-blind study was likely due to the shorter observation period and the lower amount of hydrogen compared to those of the open-label trial.

6.7.3 Side Effects of Radiation Therapy on Liver Cancer

A randomized placebo-controlled study of hydrogen water in 49 patients receiving radiation therapy for malignant liver tumors was performed by Kang et al. [39]. Hydrogen suppressed the elevation of total hydroperoxide levels, maintained serum antioxidant capacity, and improved the QOL scores. In particular, hydrogen efficiently prevented loss of appetite. Although the patients were randomly assigned to the hydrogen and placebo groups, the study could not be completely blinded because hydrogen was produced with a metallic magnesium stick, which generated hydrogen bubbles.

6.7.4 Brainstem Infarction

Edaravone is the only one antioxidant drug which has been approved to treat stroke. Ono et al. intravenously administered hydrogen along with edaravone in eight patients with acute brain stem infarction and compared MRI indices of 26 patients who received edaravone alone [13]. The MRI indices were all much more improved with the combined infusion of edaravone and hydrogen.

No adverse effect of hydrogen has been documented in the seven human diseases described above. Among the seven diseases, the most prominent effect was observed in subjects with metabolic syndrome, who consumed 1.5–2.0 L of hydrogen water per day. The amount of hydrogen water may be a critical parameter that determines clinical outcome. It is also interesting to note that lipid and glucose metabolisms were analyzed in three studies and all showed favorable responses to hydrogen [29, 32, 65]. This suggests a key regulatory role of hydrogen in the body fat metabolism and glucose metabolism.

6.8 Barriers in Studies on Biological Effects of Hydrogen

Although research on the biological effects of hydrogen is vigorous, there is no substantive progress in those studies after several years. Even some basic academic issues do not have clear answers.

There are two major problems or mysteries in the effects of hydrogen. First, dose–effect relationship was still unclear. No matter animal experiments or clinical observation, the dosage of hydrogen is small, but the effects are significant. Second, a large amount of hydrogen can be produced by gut bacteria in human and animals, but why the increase of such a small amount of hydrogen can generate such significant effects? In addition, other issues, such as the molecular basis of hydrogen, are the best way to use hydrogen for different diseases; the dose, frequency, etc. still need more research.

Although the biological effect of hydrogen has been demonstrated in more than 60 animal disease models and 6 human diseases, only two kinds of diseases, including cerebral infarction and metabolic syndrome, have both animal and clinical studies. Hydrogen has no side effects, which is a significant advantage to carry out clinical research. Some human diseases, including Parkinson's disease, are carried out which have achieved some good results. We believe that hydrogen will have certain therapeutic value for more human diseases. Of course, exploring the mechanism of hydrogen is very worthy and may be full of hardship.

References

1. Ohsawa I, Ishikawa M, Takahashi K, et al. Hydrogen acts as a therapeutic antioxidant by selectively reducing cytotoxic oxygen radicals. Nat Med. 2007;13:688–94.
2. Ohta S. Molecular hydrogen as a preventive and therapeutic medical gas: initiation, development and potential of hydrogen medicine. Pharmacol Ther. 2014. Oct;144(1):1-11.
3. Dole M, Wilson FR, Fife WP. Hyperbaric hydrogen therapy: a possible treatment for cancer. Science. 1975;190:152–4.
4. Gharib B, Hanna S, Abdallahi OM, Lepidi H, Gardette B, De Reggi M. Anti-inflammatory properties of molecular hydrogen: investigation on parasite-induced liver inflammation. C R Acad Sci III. 2001;324:719–24.

5. Matchett GA, Fathali N, Hasegawa Y, et al. Hydrogen gas is ineffective in moderate and severe neonatal hypoxia–ischemia rat models. Brain Res. 2009;1259:90–7.
6. Fujita R, Tanaka Y, Saihara Y, Yamakita M, Ando D, Koyama K. Effect of molecular hydrogen saturated alkaline electrolyzed water on disuse muscle atrophy in gastrocnemius muscle. J Phys Anthropol. 2011;30:195–201.
7. Powers SK, Smuder AJ, Judge AR. Oxidative stress and disuse muscle atrophy: cause or consequence? Curr Opin Clin Nutr Metab Care. 2012;15:240–5.
8. Kajiya M, Sato K, Silva MJ, et al. Hydrogen from intestinal bacteria is protective for Concanavalin A-induced hepatitis. Biochem Biophys Res Commun. 2009;386:316–21.
9. Cai J, Kang Z, Liu WW, et al. Hydrogen therapy reduces apoptosis in neonatal hypoxia-ischemia rat model. Neurosci Lett. 2008;441:167–72.
10. Kubota M, Shimmura S, Kubota S, et al. Hydrogen and N-acetyl-L-cysteine rescue oxidative stress-induced angiogenesis in a mouse corneal alkali-burn model. Invest Ophthalmol Vis Sci. 2011;52:427–33.
11. Li J, Wang C, Zhang JH, Cai JM, Cao YP, Sun XJ. Hydrogen-rich saline improves memory function in a rat model of amyloid-beta-induced Alzheimer's disease by reduction of oxidative stress. Brain Res. 2010;1328:152–61.
12. Wang C, Li J, Liu Q, et al. Hydrogen-rich saline reduces oxidative stress and inflammation by inhibit of JNK and NF-kappaB activation in a rat model of amyloid-beta-induced Alzheimer's disease. Neurosci Lett. 2011;491:127–32.
13. Ono H, Nishijima Y, Adachi N, et al. Improved brain MRI indices in the acute brain stem infarct sites treated with hydroxyl radical scavengers, Edaravone and hydrogen, as compared to Edaravone alone. A non-controlled study. Med Gas Res. 2011;1:12.
14. Liu Y, Liu W, Sun X, et al. Hydrogen saline offers neuroprotection by reducing oxidative stress in a focal cerebral ischemia-reperfusion rat model. Med Gas Res. 2011;1:15.
15. Fujita K, Seike T, Yutsudo N, et al. Hydrogen in drinking water reduces dopaminergic neuronal loss in the 1-methyl-4-phenyl-1,2,3,6-tetrahydropyridine mouse model of Parkinson's disease. PloS one. 2009;4:e7247.
16. Fu Y, Ito M, Fujita Y, et al. Molecular hydrogen is protective against 6-hydroxydopamine-induced nigrostriatal degeneration in a rat model of Parkinson's disease. Neurosci Lett. 2009;453:81–5.
17. Yoritaka A, Takanashi M, Hirayama M, Nakahara T, Ohta S, Hattori N. Pilot study of H(2) therapy in Parkinson's disease: a randomized double-blind placebo-controlled trial. Mov Disord. 2013;28:836–9.
18. Shen L, Wang J, Liu K, et al. Hydrogen-rich saline is cerebroprotective in a rat model of deep hypothermic circulatory arrest. Neurochem Res. 2011;36:1501–11.
19. Jiao L, Zhang J, Li Z, Liu H, Chen Y, Xu S. Edaravone alleviates delayed neuronal death and long-dated cognitive dysfunction of hippocampus after transient focal ischemia in Wistar rat brains. Neurosci. 2011;182:177–83.
20. Zhan Y, Chen C, Suzuki H, Hu Q, Zhi X, Zhang JH. Hydrogen gas ameliorates oxidative stress in early brain injury after subarachnoid hemorrhage in rats. Crit Care Med. 2012;40:1291–6.
21. Hong Y, Guo S, Chen S, Sun C, Zhang J, Sun X. Beneficial effect of hydrogen-rich saline on cerebral vasospasm after experimental subarachnoid hemorrhage in rats. J Neurosci Res. 2012;90:1670–80.
22. Takeuchi S, Wada K, Nagatani K, Osada H, Otani N, Nawashiro H. Hydrogen may inhibit collagen-induced platelet aggregation: an ex vivo and in vivo study. Intern Med. 2012;51:1309–13.
23. Chen CH, Manaenko A, Zhan Y, et al. Hydrogen gas reduced acute hyperglycemia-enhanced hemorrhagic transformation in a focal ischemia rat model. Neuroscience. 2010;169:402–14.
24. Schapira AH. Mitochondria in the aetiology and pathogenesis of Parkinson's disease. Lancet Neurol. 2008;7:97–109.
25. Nakayama M, Nakano H, Hamada H, Itami N, Nakazawa R, Ito S. A novel bioactive haemodialysis system using dissolved dihydrogen (H2) produced by water electrolysis: a clinical trial. Nephrol Dial Transplant. 2010;25:3026–33.

26. Jucker M, Walker LC. Pathogenic protein seeding in Alzheimer disease and other neurodegenerative disorders. Ann Neurol. 2011;70:532–40.
27. Nagata K, Nakashima-Kamimura N, Mikami T, Ohsawa I, Ohta S. Consumption of molecular hydrogen prevents the stress-induced impairments in hippocampus-dependent learning tasks during chronic physical restraint in mice. Neuropsychopharmacology. 2009;34:501–8.
28. Gu Y, Huang CS, Inoue T, et al. Drinking hydrogen water ameliorated cognitive impairment in senescence-accelerated mice. J Clin Biochem Nutr. 2010;46:269–76.
29. Kajiyama S, Hasegawa G, Asano M, et al. Supplementation of hydrogen-rich water improves lipid and glucose metabolism in patients with type 2 diabetes or impaired glucose tolerance. Nutr Res. 2008;28:137–43.
30. Li Y, Hamasaki T, Nakamichi N, et al. Suppressive effects of electrolyzed reduced water on alloxan-induced apoptosis and type 1 diabetes mellitus. Cytotechnology. 2011;63:119–31.
31. Kamimura N, Nishimaki K, Ohsawa I, Ohta S. Molecular hydrogen improves obesity and diabetes by inducing hepatic FGF21 and stimulating energy metabolism in db/db mice. Obesity. 2011;19:1396–403.
32. Nakao A, Toyoda Y, Sharma P, Evans M, Guthrie N. Effectiveness of hydrogen rich water on antioxidant status of subjects with potential metabolic syndrome-an open label pilot study. J Clin Biochem Nutr. 2010;46:140–9.
33. Doyle MR, Davis SJ, Bastow RM, et al. The ELF4 gene controls circadian rhythms and flowering time in Arabidopsis thaliana. Nature. 2002;419:74–7.
34. Fukuda K, Asoh S, Ishikawa M, Yamamoto Y, Ohsawa I, Ohta S. Inhalation of hydrogen gas suppresses hepatic injury caused by ischemia/reperfusion through reducing oxidative stress. Biochem Biophys Res Commun. 2007;361:670–4.
35. Tsai CF, Hsu YW, Chen WK, et al. Hepatoprotective effect of electrolyzed reduced water against carbon tetrachloride-induced liver damage in mice. Food Chem Toxicol. 2009;47:2031–6.
36. Sun H, Chen L, Zhou W, et al. The protective role of hydrogen-rich saline in experimental liver injury in mice. J Hepatol. 2011;54:471–80.
37. Liu Q, Shen WF, Sun HY, et al. Hydrogen-rich saline protects against liver injury in rats with obstructive jaundice. Liver Int. 2010;30:958–68.
38. Kawai D, Takaki A, Nakatsuka A, et al. Hydrogen-rich water prevents progression of nonalcoholic steatohepatitis and accompanying hepatocarcinogenesis in mice. Hepatology. 2012;56:912–21.
39. Kang KM, Kang YN, Choi IB, et al. Effects of drinking hydrogen-rich water on the quality of life of patients treated with radiotherapy for liver tumors. Med Gas Res. 2011;1:11.
40. Oharazawa H, Igarashi T, Yokota T, et al. Protection of the retina by rapid diffusion of hydrogen: administration of hydrogen-loaded eye drops in retinal ischemia-reperfusion injury. Invest Ophthalmol Vis Sci. 2010;51:487–92.
41. Ono H, Nishijima Y, Adachi N, et al. Hydrogen(H2) treatment for acute erythymatous skin diseases. A report of 4 patients with safety data and a non-controlled feasibility study with H2 concentration measurement on two volunteers. Med Gas Res. 2012;2:14.
42. Mei K, Zhao S, Qian L, Li B, Ni J, Cai J. Hydrogen protects rats from dermatitis caused by local radiation. J Dermatolog Treat. 2014 Apr;25(2):182-8.
43. Murphy BA, Gilbert J. Dysphagia in head and neck cancer patients treated with radiation: assessment, sequelae, and rehabilitation. Semin Radiat Oncol. 2009;19:35–42.
44. Hymes SR, Strom EA, Fife C. Radiation dermatitis: clinical presentation, pathophysiology, and treatment 2006. J the Am Acad Dermatol. 2006;54:28–46.
45. Guo Z, Zhou B, Li W, Sun X, Luo D. Hydrogen-rich saline protects against ultraviolet B radiation injury in rats. J Biomed Res. 2012 Sep;26:365–71.
46. Ignacio RM, Yoon Y-S, Sajo MEJ, et al. The balneotherapy effect of hydrogen reduced water on UVB-mediated skin injury in hairless mice. Mol Cell Toxicol. 2013;9:15–21.
47. Shin MH, Park R, Nojima H, Kim HC, Kim YK, Chung JH. Atomic hydrogen surrounded by water molecules, H(H2O)m, modulates basal and UV-induced gene expressions in human skin in vivo. PloS one. 2013 Apr 24;8(4):e61696.

48. Nakamura T, Henson PM, Murphy RC. Occurrence of oxidized metabolites of arachidonic acid esterified to phospholipids in murine lung tissue. Anal Biochem. 1998;262:23–32.
49. Yoshimi N, Ikura Y, Sugama Y, et al. Oxidized phosphatidylcholine in alveolar macrophages in idiopathic interstitial pneumonias. Lung. 2005;183:109–21.
50. Whidden MA, McClung JM, Falk DJ, et al. Xanthine oxidase contributes to mechanical ventilation-induced diaphragmatic oxidative stress and contractile dysfunction. J Appl Physiol. 2009;106:385–94.
51. Bem RA, van Woensel JB, Bos AP, et al. Mechanical ventilation enhances lung inflammation and caspase activity in a model of mouse pneumovirus infection. Am J Physiol Lung Cell Mol Physiol. 2009;296:L46–56.
52. Huang CS, Kawamura T, Lee S, et al. Hydrogen inhalation ameliorates ventilator-induced lung injury. Crit Care. 2010;14:R234.
53. Yao L, Xu F, Luo C, et al. [Protective effect of hydrogen against hyperoxia-induced type II alveolar epithelial cell injury]. Nan Fang Yi Ke Da Xue Xue Bao. 2013;33:193–6.
54. Mao YF, Zheng XF, Cai JM, et al. Hydrogen-rich saline reduces lung injury induced by intestinal ischemia/reperfusion in rats. Biochem Biophys Res Commun. 2009;381:602–5.
55. Para AE, Bezjak A, Yeung IW, Van Dyk J, Hill RP. Effects of genistein following fractionated lung irradiation in mice. Radiother Oncol. 2009;92:500–10.
56. Deeg HJ. Acute and delayed toxicities of total body irradiation. Seattle Marrow Transplant Team. Int J Radiat Oncol Biol Phys. 1983;9:1933–9.
57. Beyzadeoglu M, Oysul K, Dirican B, et al. Effect of dose-rate and lung dose in total body irradiation on interstitial pneumonitis after bone marrow transplantation. Tohoku J Exp Med. 2004;202:255–63.
58. Sampath S, Schultheiss TE, Wong J. Dose response and factors related to interstitial pneumonitis after bone marrow transplant. Int J Radiat Oncol Biol Phys. 2005;63:876–84.
59. Chuai Y, Zhao L, Ni J, et al. A possible prevention strategy of radiation pneumonitis: combine radiotherapy with aerosol inhalation of hydrogen-rich solution. Med Sci Monit. 2011 Apr;17(4):HY1-4.
60. Terasaki Y, Ohsawa I, Terasaki M, et al. Hydrogen therapy attenuates irradiation-induced lung damage by reducing oxidative stress. Am J Physiol Lung Cell Mol Physiol. 2011 Oct;301:L415–26.
61. Ishibashi T, Sato B, Rikitake M, et al. Consumption of water containing a high concentration of molecular hydrogen reduces oxidative stress and disease activity in patients with rheumatoid arthritis: an open-label pilot study. Med Gas Res. 2012;2:27.
62. Ishibashi T, Sato B, Shibata S, et al. Therapeutic efficacy of infused molecular hydrogen in saline on rheumatoid arthritis: a randomized, double-blind, placebo-controlled pilot study. Int Immunopharmacol. 2014;21:468–73.
63. He J, Xiong S, Zhang J, et al. Protective effects of hydrogen-rich saline on ulcerative colitis rat model. J Surg Res. 2013;185:174–81.
64. Nakayama M, Kabayama S, Nakano H, et al. Biological effects of electrolyzed water in hemodialysis. Nephron Clin Pract. 2009;112:c9–15.
65. Ito M, Ibi T, Sahashi K, Ichihara M, Ito M, Ohno K. Open-label trial and randomized, double-blind, placebo-controlled, crossover trial of hydrogen-enriched water for mitochondrial and inflammatory myopathies. Med Gas Res. 2011;1:24.

Chapter 7
Methods of Hydrogen Application

Liren Qian, Jianliang Shen and Xuejun Sun

Abstract Hydrogen gas has come to the forefront of therapeutic medical gas research. Recent basic and clinical studies have revealed that hydrogen gas is an important physiological regulatory factor, with antioxidant, anti-inflammatory, and antiapoptotic protective effects on cells and organs. Since hydrogen gas is explosive, the methods of hydrogen application become an important issue for researchers. Currently, the methods of hydrogen application are developing rapidly, including inhalation of hydrogen gas, oral hydrogen water, hydrogen saline injection, eye drops, skin smear and bathing, hydrogen gas injection, etc. In this chapter, various methods of hydrogen application are discussed with particular attention to the experimental models and human beings.

Keywords Hydrogen · Method · Hydrogen-rich water · Electrolytic water · Metallic magnesium · Electric acupuncture needle

If hydrogen can indeed cure diseases, how to make it easier to use become an issue many researchers concern inevitably. Up to now, ways of hydrogen application on different diseases include inhalation of hydrogen gas, hydrogen water, hydrogen-rich saline injection, hydrogen-rich water transdermal diffusion and inducing colonic bacteria to produce hydrogen by food or drugs, etc [1–7]. Electrolysis of water, as a kind of functional water, has been used for many years in the world. Hydrogen-rich water, as a novel product of functional water, is developing rapidly in Japan, Korea, Southeast Asia, and other regions. Hydrogen-rich saline injection is one of the most outstanding contributions in this field with obvious advantages. As a kind of clinical treatment, hydrogen-rich saline injection still needs further study in basic and clinical researches. But as a kind of very promising treatment, hydrogen-rich saline injection is bound to become the most promising way in clinical application of hydrogen [8]. Diffusing hydrogen through the skin, local injection of hydrogen

X. Sun (✉)
Department of Navy Aeromedicine, Second Military Medical University, Shanghai, China
e-mail: sunxjk@hotmail.com

L. Qian · J. Shen
Department of Hematology, Navy General Hospital, Fucheng Road, Beijing, China

X. Sun et al. (eds.), *Hydrogen Molecular Biology and Medicine,*
DOI 10.1007/978-94-017-9691-0_7

gas, drug-induced and food-induced hydrogen by colon bacteria have great value, which are also worthy to study.

7.1 Inhalation

A great deal of human trials have been done using hydrogen gas inhalation in diving medicine. This proved the safety of hydrogen gas and provided a number of valuable experiences on treating diseases by hydrogen gas. As compared to other gases, such as oxygen, helium, carbon dioxide, etc., it is natural to envision the use of hydrogen gas by inhalation. But hydrogen gas is explosive, which has to be overcome. Under pure oxygen gas environment, it is not flammable or explosive when the hydrogen gas concentration is less than 4 % or over 95 %. Hydrogen gas is flammable when its concentration is between 4 and 75 % in air. But whether hydrogen gas has biological effects at such concentration needs to be determined. Fortunately, experimental evidence in 2007 suggests that 1 or 2 % hydrogen gas inhalation for 35 min is effective in treating cerebral ischemia-reperfusion injury in animal models. Subsequently, a large number of studies have shown that a small concentration of hydrogen gas inhalation has therapeutic effects on a variety of diseases. These studies will provide a foundation for treating diseases by hydrogen gas inhalation. An important issue for treating diseases by hydrogen gas inhalation is its safety. Although breathing low concentration of hydrogen gas is relatively a safe method, localized high concentration of hydrogen gas is impossible to avoid which requires strict safety operations in using hydrogen gas. Concentration and ignition point of hydrogen gas are the two most dangerous factors we need to consider. It is impossible to be explosive as long as the concentration of hydrogen gas is less than 4 % or over 95 % in pure oxygen environment, or over 75 % in air. If the temperature does not exceed 500 °C, hydrogen gas would not burn. Although 500 °C is high, it is easy to reach. Even static spark can achieve such a temperature. Prevention of static spark is of great importance during operation.

The two most important parameters are the total volume of the mixture and the concentration of hydrogen when considering the destruction by hydrogen gas explosion. If the volume of hydrogen gas is twice that of oxygen, it will produce explosive effects. The farther away from the ratio, the smaller explosive effects the mixture would produce. What is more important is the total volume of the gas mixture. Great destruction cannot be caused, even burn or explosion happens, if the total volume of the gas mixture is very small. According to these characteristics, medical equipment for hydrogen gas inhalation has been invented.

Another disadvantage of hydrogen gas inhalation is the uncertainty of the dosage of hydrogen gas uptake. The dosage of hydrogen gas inhaled can be affected by many factors. Hydrogen gas enters human body through respiration and blood circulation. All factors affect the patients' respiration and blood circulation and can affect the actual extent of absorption of hydrogen gas. Thus, the cardiopulmonary function of patients is an important factor. The difference of cardiopulmonary function among different patients causes the difference of actual dosage of hydrogen gas uptake among different patients.

External stimuli, such as exercise and mental stimulation, resulting in sympathetic excitement and respiratory and circulatory hyperactivity, can affect the uptake of hydrogen gas. The concentration of carbon dioxide is an important factor. Carbon dioxide causes vasoconstriction, resulting in reduced blood flow in tissues (except brain and heart). If you use a respiratory mask, increased dead space would increase the concentration of carbon dioxide, resulting in changes described above. Ambient temperature and humidity also affect respiratory and circulatory functions. Besides, ambient temperature affects the skin temperature. Increased skin temperature accelerates the release of hydrogen gas from the skin. Therefore, in the study carried out in hydrogen inhalation, the quality of respiratory gas, ambient temperature, etc. should be well controlled in order to improve the comparability between different individuals.

7.2 Hydrogen Water

7.2.1 Oral Hydrogen-Rich Water

Drinking hydrogen-rich water is the most widely applied method for uptake of hydrogen. Oral hydrogen-rich water is the most common form of health products. Oral hydrogen-rich water can be produced by electrolysis of water, dissolving hydrogen gas into water, and reaction of metallic magnesium with water, etc., which is usually used in experimental studies. It is easy to control dosage of hydrogen gas in human experiments by oral hydrogen-rich water, such as drinking fixed volume of hydrogen water within fixed time, etc. However, it is difficult to control dosage of hydrogen gas in animal experiments, because oral hydrogen-rich water is always given optionally in animal experiments. This is bound to bring some issues that affect the stability of experimental results because of multivariate absorption process, errors in quantity, volatilization of hydrogen gas from the vessel, etc. To solve these problems, some researchers limit the amount of oral hydrogen water by giving fixed volume of hydrogen water within fixed time to animals.

7.2.2 Electrolytic Water

Hydrogen-rich water was firstly used as electrolytic water in Japan for healthcare. After decades of application, Japanese government's Ministry of Health Labor approved a new generation of medical devices for electrolytic water which is called electrolyzer. Main internal elements of the device include electrode and ionic membrane. Commercially available electrolytic water manufacturing apparatus is called water ionizer or electrolytic water generator. Electrolytic water is the product of water containing salt (such as sodium chloride) by electrolysis. Electrolytic water is neutral in itself, but two kinds of water can be generated by adding other ions or by adding semipermeable membrane: alkaline water and acidic water. Sodium hydroxide, hypochlorous acid, and sodium hypochlorite can be generated by electrolyzing water containing sodium chloride. However, only hydroxide ion, hydrogen gas, oxygen gas, and hydrogen ion were generated by electrolyzing pure water. Under certain conditions, acidic electrolytic water exerted bactericidal effects. Some advertisements claimed that alkaline electrolytic water can "neutralize acidic physique" of human body, but when alkaline water reaches the stomach, it will be with a strong acidic gastric acid and becomes acidic. Therefore, adjusting the body's pH by alkaline water is lack of scientific basis.

In 1931, the world's first water ionizer was invented in Japan. In 1932, Japanese began to study the impact of electrolytic water on flora and fauna. In 1954, civil electrolytic device was successfully invented and agricultural electrolytic device was on sale. In 1960, electrolytic device was used for drinking water by Japanese and become a medical device. In 1966, Japanese No.1 medical device certification (medical apparatus for producing electrolytic water) for drinking water was issued. In 1979, Japanese certification for new generation of device was passed, which could continuously produce drinking water. In 1994, a research committee for electrolytic water was established in Japanese Ministry of Health. In 1994, the Japanese Cancer Prevention Center released a report, "Free radicals are carcinogenic incentives." They also confirmed that the electrolytic water can really scavenge free radicals in human body. Although the study of electrolytic water has a very long history, but the theory of its therapeutic effects on diseases is not very convincing. In industrial field, it is generally believed that it can treat disease because it is a kind of alkaline water with negative potential and small molecules. But these theories lack rigorous scientific evidences and are more of speculations and assumptions. Recently, in academic field, it is generally believed that it exerts therapy effects by hydrogen gas contained in electrolytic water. Therapeutic effects of hydrogen gas on different diseases by its antioxidative ability have been proved. Electrolytic water contains certain concentration of hydrogen gas which reaches the concentration exerting its antioxidative effects. If hydrogen gas in electrolytic water evaporates, the negative potential of electrolytic water disappears, leading to the disappearance of its function. Although negative potential is an indirect indicator of the concentration of dissolved hydrogen gas, it is not associated with its therapeutic effects.

7.2.3 Hydrogen-Rich Water Produced by Metallic Magnesium

Hydrogen gas and magnesium hydrate can be produced slowly by reaction between magnesium and water at room temperature which is convenient for use. Metals such as iron, aluminum, and magnesium can react with acidic water and water to generate hydrogen gas, respectively. However, they are not suitable for producing oral hydrogen-rich water because of the taste, reaction rate, toxicity, and other reasons. For convenience and disinfection, some materials such as tourmaline and nanoplatinum are added into magnesium. But magnesium is the core material and hydrogen gas plays the key role in its therapeutic effects. The main problem is the stability of hydrogen gas generated in the water. At the beginning, its effects may be desirable. But after a period of time, concentration of hydrogen gas generated will reduce gradually. Therefore, in clinical and animal studies, scheduled replacement of the product is essential. Magnesium ion is an essential metal element in human body, which is harmless to human body. However, patients with kidney failure using it must be under medical guidance.

7.2.4 Hydrogen Saturated Water

Uptake of hydrogen gas by hydrogen saturated water is widely recognized as the best way which is mainly because hydrogen saturated water has the most experimental data. Since 2008, over 100 scientific studies have been reported [9–11]. Therefore, biological effects of hydrogen saturated water are with the most certainty. Preparation of hydrogen saturated water includes four methods: aeration, high pressure, membrane separation, and the electrolytic water. During the process of hydrogen saturated water production and preservation, the most crucial technology is not how to dissolve hydrogen gas but how to avoid leakage of hydrogen gas from a container. Many packaging methods that prevent leakage of other gases cannot prevent leakage of hydrogen gas because hydrogen molecule is very small. For example, the packaging methods limit the release of carbon dioxide from the carbonated beverage package and cannot limit the release of hydrogen gas. Packaging method is the core technology in the field which is also the most difficult technical problem to solve during product development. Package is made of metal aluminum by most Japanese hydrogen saturated water suppliers, including flexible packaging aluminum bags or aluminum cans, which make hydrogen gas stable for more than 6 months. Han Hydrogen Technology Co. Ltd. of Taiwan chose glass package and achieved favorable results. Although studies on the biological effects of hydrogen gas have been carried out for many years, the development of hydrogen saturated water products is very slow because of the limitation in packaging materials and packaging methods.

Filling hydrogen gas into water to produce hydrogen saturated water is the most classic way which was thought may not be an effective way for preparing the desired concentration of hydrogen gas. However, recent studies found that even with such a simple method, hydrogen saturated water with therapeutic effects can also be produced by simply blowing hydrogen gas into the water for 10 min. We can produce a bottle of health water which is good for health by this simple method.

7.3 Hydrogen Saline Injection

There is no fundamental difference between the hydrogen-rich saline and oral hydrogen-rich water. Both of them are prepared by dissolving hydrogen gas into solution. However, for clinical application, sterile injection is needed. Hydrogen-rich saline injection has special advantages. This is a kind of promising clinical technique. This kind of method is not influenced by patients' own factors. The dose of hydrogen gas uptake can be accurately controlled. Hydrogen-rich saline injection has been carried out in various types of organ ischemia-reperfusion injuries, inflammatory diseases, arteriosclerosis, hypertension, liver damage, diabetes, and other diseases [11–14]. A new technique for producing hydrogen-rich saline has been recently developed in Japan. Normal saline can be turned into hydrogen-rich saturated saline when the package is not opened by this technology. The diffusion ability of hydrogen gas plays a key role in this technology. By immersing ordinary polyethylene bags of physiological saline into hydrogen saturated solution for 48 h, hydrogen gas diffuses through polyethylene material into physiological saline. The obtained hydrogen-rich saline by this kind of method has been applied on clinical trials in Japan, which has been proved available [15].

7.4 Other Methods of Hydrogen Application

Besides the three methods used above, some unique methods are also used, although these methods are seldom used.

7.4.1 Eyedrops

Hydrogen eye drops have been proved effective on retina ischemia-reperfusion injury. The study showed potential and broad application prospects of hydrogen eye drops [16].

7.4.2 Skin Smear and Bathing

Due to the strong diffusion capacity of hydrogen gas, it can be absorbed through the skin. Some researchers treated skin injury caused by inflammation by combination of drinking hydrogen-rich water with local application of hydrogen-rich water [17]. It showed significant therapeutic effects on most patients in 1 or 2 weeks. Hydrogen gas can also be used by putting part of the body in a closed hydrogen gas atmosphere, although this kind of method has not been reported.

7.4.3 Hydrogen Gas Injection

Researchers from Zhongshan Medical University have done research comparing the therapeutic effects of intraperitoneal injection of hydrogen gas and hydrogen-rich saline on cerebral ischemia. More than 60 times volume of hydrogen gas can be obtained by injecting hydrogen gas directly than injecting the same volume of hydrogen-rich saline because of the lower solubility of hydrogen gas. However, injection of hydrogen gas may cause emphysema and infection; the safety and effectiveness of this method needs more research.

7.4.4 Hydrogen Gas Produced by Intestinal Bacteria

Those drugs and food that cannot be absorbed by small intestine are transported to colon, where hydrogen gas can be produced by the bacteria. Some researchers have demonstrated that oral acarbose, modified starch, milk, curcumin, and lactulose can promote hydrogen gas generated in human body [18, 19]. Other ingredients of food that can promote bacteria-generating hydrogen gas in colon include cottonseed sugar, lactose, sorbitol, mannitol, chitosan, soluble fiber, etc [19]. A huge volume of hydrogen gas can be induced by intestinal bacteria. Although a part of it can be used by some other bacteria, such as methane bacteria, much can be absorbed into the blood circulation through the intestine mucosa, and transported to other organs to play a role in the treatment of diseases. This mean has been studied for some time, but the effect is still uncertain.

7.4.5 Hydrogen Gas Produced by Oral Drugs

Magnesium was used as a kind of drug for gastritis. Hydrogen gas can be produced by metal magnesium when reacted with gastric acid. There is still no experimental evidence that this method can treat diseases. Statins can inhibit methane bacteria, which consume hydrogen gas to produce methane. Therefore, statins indirectly increase intestinal hydrogen gas content. We speculated that the increase of hydrogen gas content in the intestinal tract is one of the reasons while statins can protect heart.

7.4.6 Electric Acupuncture Needle and Direct Current

Tissues can be electrolyzed by electric acupuncture needle. At the anticathode, due to the loss of electron, electrolyzed water produces a certain amount of oxygen gas. Because of the complexity of the composition of tissue fluid, various kinds of reactive oxygen species may be produced at the anticathode, which may induce certain oxidative damage. At the cathode, hydrogen ion turns into hydrogen atom when it acquires electron, which finally turns into hydrogen gas. Because the concentration of hydrogen gas in tissues increased when using electric acupuncture needle treating disease, we have to consider the role of hydrogen gas in acupuncture treatment. Many studies have demonstrated that acupuncture treatment has antioxidative and anti-inflammatory effects, which happens to be the biological effects of hydrogen gas.

If we can prove that treating diseases by electric acupuncture needle is through hydrogen gas, it will be valuable not only in the study of hydrogen gas but also in the study of electric acupuncture needle.

References

1. Ohsawa I, Ishikawa M, Takahashi K, Watanabe M, Nishimaki K, Yamagata K, Katsura K, Katayama Y, Asoh S, Ohta S. Hydrogen acts as a therapeutic antioxidant by selectively reducing cytotoxic oxygen radicals. Nat Med. 2007;13(6):688–94.
2. Lee MY, Kim YK, Ryoo KK, Lee YB, Park EJ. Electrolyzed-reduced water protects against oxidative damage to DNA, RNA, and protein. Appl Biochem Biotechnol. 2006;135(2):133–44.
3. Ohsawa I, Nishimaki K, Yamagata K, Ishikawa M, Ohta S. Consumption of hydrogen water prevents atherosclerosis in apolipoprotein E knockout mice. Biochem Biophys Res Commun. 2008;377(4):1195–8.
4. Cai J, Kang Z, Liu K, Liu W, Li R, Zhang JH, Luo X, Sun X. Neuroprotective effects of hydrogen saline in neonatal hypoxia-ischemia rat model. Brain Res. 2009;1256:129–37.
5. Kajiya M, Sato K, Silva MJ, Ouhara K, Do PM, Shanmugam KT, Kawai T. Hydrogen from intestinal bacteria is protective for Concanavalin A-induced hepatitis. Biochem Biophys Res Commun. 2009;386(2):316–21.
6. Huang G, Zhou J, Zhan W, Xiong Y, Hu C, Li X, Li X, Li Y, Liao X. The neuroprotective effects of intraperitoneal injection of hydrogen in rabbits with cardiac arrest. Resuscitation. 2013;84(5):690–5.
7. Ball R. Biochemical decompression of hydrogen by naturally occurring bacterial flora in pigs: what are the implications for human hydrogen diving? Undersea Hyperb Med. 2001;28(2):55–6. (Journal of the Undersea and Hyperbaric Medical Society, Inc.)
8. Kang KM, Kang YN, Choi IB, Gu Y, Kawamura T, Toyoda Y, Nakao A. Effects of drinking hydrogen-rich water on the quality of life of patients treated with radiotherapy for liver tumors. Med Gas Res. 2011;1(1):11.
9. Qian L, Shen J, Chuai Y, Cai J. Hydrogen as a new class of radioprotective agent. Int J Biol Sci. 2013;9(9):887–94.
10. Ohta S. Molecular hydrogen as a preventive and therapeutic medical gas: initiation, development and potential of hydrogen medicine. Pharmacol Ther. 2014;144:1–11.
11. Ohno K, Ito M, Ichihara M, Ito M. Molecular hydrogen as an emerging therapeutic medical gas for neurodegenerative and other diseases. Oxid Med Cell Longev. 2012;2012:353152.
12. Qian L, Mei K, Shen J, Cai J. Administration of hydrogen-rich saline protects mice from lethal acute graft-versus-host disease (aGVHD). Transplantation. 2013;95(5):658–62.
13. Qian L, Cao F, Cui J, Huang Y, Zhou X, Liu S, Cai J. Radioprotective effect of hydrogen in cultured cells and mice. Free Radic Res. 2010;44(3):275–82.
14. Qian L, Cao F, Cui J, Wang Y, Huang Y, Chuai Y, Zaho L, Jiang H, Cai J. The potential cardioprotective effects of hydrogen in irradiated mice. J Radiat Res. 2010;51(6):741–7.
15. Ono H, Nishijima Y, Adachi N, Sakamoto M, Kudo Y, Nakazawa J, Kaneko K, Nakao A. Hydrogen(H2) treatment for acute erythymatous skin diseases. A report of 4 patients with safety data and a non-controlled feasibility study with H2 concentration measurement on two volunteers. Med Gas Res. 2012;2(1):14.
16. Oharazawa H, Igarashi T, Yokota T, Fujii H, Suzuki H, Machide M, Takahashi H, Ohta S, Ohsawa I. Protection of the retina by rapid diffusion of hydrogen: administration of

hydrogen-loaded eye drops in retinal ischemia-reperfusion injury. Invest Ophthalmol Vis Sci. 2010;51(1):487–92.
17. Ignacio RM, Yoon Y-S, Sajo MEJ, Kim C-S, Kim D-H, Kim S-K, Lee K-J. The balneotherapy effect of hydrogen reduced water on UVB-mediated skin injury in hairless mice. Mol Cell Toxicol. 2013;9(1):15–21.
18. Chen X, Zuo Q, Hai Y, Sun XJ. Lactulose: an indirect antioxidant ameliorating inflammatory bowel disease by increasing hydrogen production. Med Hypotheses. 2011;76(3):325–7.
19. Liu S, Sun Q, Tao H, Sun X. Oral administration of mannitol may be an effective treatment for ischemia-reperfusion injury. Med Hypotheses. 2010;75(6):620–2.

Chapter 8
Future Directions in Hydrogen Studies

Xiao Chen, Xuejun Sun and Shigeo Ohta

Abstract Although the biological effect of hydrogen was first reported in 1975, it did not draw much attention. Since the protective effects of small amount of hydrogen on oxidative stress were revealed in 2007, recent years have witnessed a tremendous development of hydrogen biology. Hydrogen has been proved effective in various diseases including ischemia/reperfusion injuries, diabetes, pancreatitis, ulcerative colitis, and so on. However, the underlying mechanism is still not clear. Neutralizing toxic oxygen species only explains part of the phenomena and is now being challenged by increasingly accumulated evidence. The future direction for hydrogen biology lies in molecular mechanism, clinical study, and in other biological systems. Besides, some other small molecules like helium and methane are worth exploring.

Keywords Hydrogen · Development · Biological effects · Future direction

Hydrogen biology studies have become a hot spot in the bioscience area. Since 2007, more than 300 papers in this area have been published in 5 years. In early times, mainly laboratories from Japan, China, and America carried out related studies. Nowadays, the team has been joined by many famous laboratories from Europe and is growing bigger and bigger. In May 2010, major scholars in this area jointly founded the magazine *Medical Gas Research* in which researches regarding hydrogen biology and medicine have accounted for more than 50% of all studies published.

Concerning medical gas study, the best-known gases are nitrogen monoxidum (NO), carbon monoxide (CO), and hydrogen sulfide (H_2S) in recent 10 years.

S. Ohta (✉)
Department of Biochemistry and Cell Biology, Institute of Development and Aging Sciences, Graduate School of Medicine, Nippon Medical School, Kawasaki, Japan
e-mail: ohta@nms.ac.jp

X. Chen
Department of Orthopedics trauma, Shanghai Changhai Hospital, Second Military Medical University, Shanghai, China

X. Sun
Department of Navy Aeromedicine, Second Military Medical University, Shanghai, China

X. Sun et al. (eds.), *Hydrogen Molecular Biology and Medicine*,
DOI 10.1007/978-94-017-9691-0_8

Whether hydrogen is able to join the family depends on the breakthroughs on the mechanisms of hydrogen biological effects. Merely focusing on the macroscopic biological effects without elaborating the mechanisms leaves no room for hydrogen biology in the bioscience area which gives much attention to the basic theory of exploration. Therefore, the researchers in hydrogen biologic effects should attach great importance to the molecular mechanisms.

In an article in 2008, I proposed that hydrogen studies might win the Nobel Prize in Physiology or Medicine. However, it requires two vital premises: hydrogen is proved clinically effective and widely popularized in clinical practice and the mechanisms of hydrogen biological effects are well elaborated. Based on the current studies in this area, major breakthroughs are not very likely in short term. I hope the dedicated young Chinese scientists could overcome the difficulties and solve the enigma.

8.1 Train of Thoughts in Early Studies

It has long been noted that hydrogen is an important metabolic substance among lower organisms. For example, some bacteria can produce hydrogen and some can use and degrade hydrogen. Nevertheless, rare studies had concerned the participation of hydrogen in normal metabolic process or diseases among higher organisms before 2007.

It is widely recognized that hydrogen is a strong reducing chemical reagent. Similar to oxygen, hydrogen has a relatively low solubility. However, with the presence of hemoglobin, with which oxygen can be combined, oxygen can be absorbed in a large amount through breathing. Therefore, for a long time, it has been defined by biologists, especially diving physicians, that hydrogen is a physiologically inert gas. Namely, hydrogen is not involved in any biological reaction in vivo with any substance. Early in diving medicine, hydrogen was breathed in under high pressure and the dissolved hydrogen increased. Scientists tried to prove that under high pressure, hydrogen could react with oxygen or free radicals in solution. Nonetheless, it has not been demonstrated due to the design of experiments. On the one hand, probably the sensitivity of the detection method was not enough. On the other hand, it did show that it was difficult for hydrogen to participate in metabolism in vivo. But as early as 1975, Dole et al. [1] demonstrated that a mixture of 2.5 % oxygen and 97.5 % hydrogen at a total pressure of 8 atm for periods up to 2 weeks could treat squamous cell carcinoma. In 2001, Gharib et al. [2] showed that breathing high-pressure hydrogen could treat parasite-induced liver inflammation. The above two studies did not draw much attention mainly because high-pressure hydrogen cannot be used in clinical practice as a usual treatment option. In radiation chemistry, it was proved that hydrogen could directly react with hydroxyl radicals in solution, which did not concern biologists either.

Looking back drives us to understand the origin of a discipline and a thought. Studies on the biological effects of hydrogen were initiated approximately 200 years ago after hydrogen was discovered. Early studies only explored the reactions of animals after breathing hydrogen. No doubt did the studies demonstrate that hydrogen has no effects on animals. With regard to the biological effects of hydrogen, diving medicine showed great interest because hydrogen was used as a diving gas and studied by many international diving medicine research institutes. As a breathing gas, safety is the priority. That is, it must not have significant influence on the human body, including breathing under extremely high pressure. Many years of diving medicine studies have proved that breathing hydrogen is very safe, which meanwhile impresses people that breathing hydrogen has no significant biological effects. As the old saying goes, one coin has two sides. Diving medicine studies have laid a very solid foundation for hydrogen study and limited its development to some extent. Nevertheless, findings in other fields broke the inertia and diving medicine studies provided valuable evidence of biological safety for hydrogen.

Although hydrogen was considered as an inert gas in diving medicine, it was used in cancer treatment in 1975 and liver parasite infection in 2001. For such a long time, little attention was paid to hydrogen. One important reason is that the hydrogen being an inert gas is deeply ingrained in people's mind. More than 20 years ago, before the application of laser Doppler techniques, based on the fact that the decrease rate of hydrogen in tissues is proportional to the blood supply of the tissue, a method employing breathing hydrogen for blood supply measurement had been widely used in many animal disease models. To measure the decrease rate of hydrogen concentration in tissues after breathing 10% hydrogen for 10–30 min could indirectly calculate the tissue blood flow. From the current perspective, the technique itself could treat certain diseases. But no one discovered the effect early, which was a great pity.

8.2 Discovery of Biological Effects of Hydrogen

In 2007, Ohsawaet al. [3] reported that inhalation of 2% hydrogen gas could selectively reduce cytotoxic oxygen radicals and markedly ameliorate cerebral ischemia/reperfusion injuries. They employed chemical and cytologic methods and demonstrated that hydrogen dissolved in solution could selectively neutralize hydroxy radicals and nitrite, which both act as important mediators in oxidative injuries. No mechanism within body could eliminate these two toxic reactive oxygen species (ROS), which serves as an important basis for many pathological processes. Afterward, they demonstrated that inhalation of 2% hydrogen gas could ameliorate hepatic [4] and myocardial [5] ischemia reperfusion injuries. Nakao et al. [6] from the Pittsburgh Transplantation Institute proved that inhalation of 2% hydrogen could ameliorate oxidative stress in transplantation-induced intestinal graft injury. For better convenience, hydrogen-saturated saline was prepared and widely used in studies. Consumption of hydrogen-rich saline prevents the stress-induced neurological impairments [7, 8] and atherosclerosis in apolipoprotein E knockout mice [9]. Hydrogen-supplemented drinking water protects cardiac [10] and renal [11] allografts from inflammation-associated deterioration. Hydrogen-rich saline could achieve similar protective effects to the hydrogen gas. But the preservation and use are much safer and more convenient. For diabetes mellitus, especially type II, hydrogen-rich water shows significant protective effects on oxidative stress injuries [12]. Domestically, Xie et al. from the Fourth Military Medical University showed that hydrogen could be used in the treatment of systemic inflammation, multiple organ failure, and acute craniocerebral injury [13–16]. Our study proved that inhalation of 2 % hydrogen was protective for neonatal hypoxia–ischemia injuries [17]. Afterward, we successfully prepared hydrogen-saturated saline and initiated collaboration with more than 40 domestic laboratories and found that saline had promising treatment effects on pain [18]; rheumatoid arthritis [19]; acute pancreatitis [20]; senile dementia [21]; chronic oxygen poisoning [22]; delayed encephalopathy after carbon monoxide intoxication [23]; liver injury [24]; spinal cord trauma [25]; chronic hypoxia [26]; peritonitis [27]; colitis [28, 29]; neonatal hypoxic–ischemic brain injury [30]; and myocardial [31, 32], lung [33], renal [34], and intestinal [35] ischemia–reperfusion injury. In particular, our study showed that early treatment with hydrogen saturated saline could significantly improve long-term neurological and neurobehavioral functions [30]. The above studies indicate that hydrogen is an ideal scavenger for ROS, especially toxic ROS with potential clinical application.

In the 5-year study of hydrogen biological effects, the team led by researchers from Japan, China, and America is growing bigger and bigger. However, the study level is not progressing correspondingly. The major reason is that this area has not attracted the attention of the frontier scientists. The world's most cutting-edge researches mainly focus on cancer stem cells, small interfering RNA, epigenetics, and central nervous system functions. The study of biological effects of hydrogen in the whole life science is only considered a partial, relatively less popular, small field.

Additionally, some special obstacles regarding study methods such as quantification and tracking are objective factors that hinder the rapid development of hydrogen biological study.

8.3 Questions on Selective Antioxidation of Hydrogen

Although the biological effects of hydrogen are confirmative and even amazing in many human diseases and animal models, the exact molecular mechanisms still remain unclear.

Due to the relatively low solubility of hydrogen, the solubility of hydrogen is by volume approximately 1.6 ml/100 ml in water and 3.0 ml/100 ml in fat. Under such a low concentration, plus the relatively low reduction ability, it is almost impossible for hydrogen to directly react with other weakly oxidizing materials such as oxygen. But there are various kinds of ROS in vivo. Some are weakly oxidizing, e.g., nitric oxide, hydrogen peroxide, and superoxide anion and some are strongly oxidizing, e.g., hydroxyl radicals and nitrite anions. Direct reactions between hydrogen and weakly oxidizing agents can hardly occur. But they can occur between hydrogen and strongly oxidizing agents such as hydroxyl radicals and nitrite anions, which has been proved with substantial evidence.

As for selectively antioxidative property of hydrogen, many questions remain unsolved. Cytoplasma is far more complicated than saline in vitro and there are tons of materials which can react with hydroxyl radicals. It cannot be deduced that hydroxyl radicals will be selectively neutralized by hydrogen in vivo from direct reactions between them in saline in vitro. From the perspective of reaction rate, reaction between hydroxyl and other biological molecules is 1000 times faster than between hydroxyl and hydrogen. Theoretically, there will be no antioxidative selectivity without a very high hydrogen concentration. In vitro studies by Jelle et al. proved that hydrogen did not directly scavenge HO• and ONOOH from the perspective of chemistry [36]. Therefore, it is more convincing if derivatives of hydroxyl radicals directly react with hydrogen. For example, nitrite anions are less active than hydroxyl radicals and have longer diffusion distance. It is more likely to encounter hydrogen and be selectively neutralized.

Only indirect in vitro evidence is not complete to prove selective antioxidativity of hydrogen and more direct in vivo evidence is needed. Some phenomena indicated that the underlying mechanisms were probably more complicated than what we already have understood. In 2007, Ohsawa et al. found that inhalation of 4% hydrogen was less protective than 2% hydrogen, which indicated that the protective effects of hydrogen were not in a linear relationship with the concentration. Mami et al. [37] demonstrated only 1/20 saturated hydrogen could effectively improve Parkinson's disease. The effects of hydrogen are more powerful than what we imagine. Some studies found certain concentrations of hydrogen could influence the concentration of hydrogen peroxide and superoxide anion. The above evidence all indicate the mechanism of selective antioxidativity is worth further exploring.

Many biochemical reactions are catalyzed by the enzyme. Whether the enzyme can affect the biological effects of hydrogen or whether hydrogen participates in biochemical reactions with the presence of enzymes catalysis remains unclear. Different from general chemical reactions limited by temperature, concentration, and so on, many biochemical reactions depend on enzyme for catalysis. The role of enzymes is to lower the activation energy and initiate chemical reactions which otherwise will not occur due to lack of conditions (temperature, concentration, etc.). There are numerous kinds of enzymes in vitro with a large quantity for catalyzing oxidation reactions. Probably some could catalyze reactions between hydrogen and weak oxidizing substances. With a small molecular weight and strong penetration power (no barrier to prevent free diffusion of hydrogen in vivo), it is very likely for these reactions to occur. Besides, hydrogenase is the key enzyme responsible for generating and metabolizing hydrogen in lower organisms. They reserve similar molecules within higher organisms. Whether these molecules keep similar functions as in lower organisms is worth exploring. We believe that the selective antioxidative property of hydrogen requires certain conditions.

Many metals such as iron and copper could absorb hydrogen. These elements are also relatively abundant within organisms. Whether hydrogen is absorbed by these elements in vivo and exerts its biological effects is worth considering. Recently, some domestic scholars proposed that the biological effects of hydrogen were catalyzed by some metal ions.

8.4 Future Directions in Studies on Biological Effects of Hydrogen

Since we published the first article in this field in 2008, Chinese scholars have contributed more than 150 articles and the total number of studies published in this area has reached more than 300. Studies conducted by Chinese researchers accounted for more than 50 %. Although we started late, we have the largest scale. The number of medical schools and research institutes participating in hydrogen studies has been increasing dramatically. Currently, there have been 16 schools and institutes involved, including School of Medicine of Fudan University, School of Medicine of Shanghai Jiaotong University, School of Medicine of Tongji University, Second Military Medical University, Fourth Military Medical University, Third Military Medical University, School of Medicine of Nanjing University, Tianjin Medical University, School of Medicine of Huaxi University, Shaanxi Normal University, China Medical University, Capital Medical University, Harbin Medical University, Soochow University, Taishan Medical College, and Wenzhou Medical College.

The number of National Natural Science Foundation of China could indirectly reflect the degree of concern of this field. Statistics showed that there have been more than 20 foundations in this field including 2 in 2009, 4 in 2010, 7 in 2011, 5 in 2012, and 11 in 2013.

Based on current publications, some points need to be noticed by Chinese researchers. First, as for the study depth, there is a significant gap compared with studies from Japan and America. Specifically, our studies mainly focused on the treatment effect's observations, and confirmed the therapeutic effects but neglected to explore the molecular mechanisms. Second, the clinical study has not been conducted yet. Although hydrogen has not been approved as medicine in clinical use, patients can still be used as a health-care means. American and Japanese scholars have taken rewarding attempts. Hydrogen-rich drinking water and hydrogen generation rods have been commercialized and used by a large quantity of people in Japan and South Asia. Its effects in diabetes, kidney failure, and atherosclerosis are explored. Clinical studies in China are lagging behind because it has not aroused real concern. However, we have many advantages such as large population and diverse diseases. As long as we choose the appropriate disease, we ought to achieve better results in clinical studies of hydrogen.

Internationally, some important problems still remain unconcerned in this field. One is endogenous hydrogen study. Administration of such unabsorbable agents as mannitol, lactulose, and many agents in Chinese herbs can induce intestinal bacteria to produce hydrogen, which could exert same therapeutic effects as breathing hydrogen theoretically. Electric acupuncture is essentially a direct stimulation. During the electric acupuncture process, the tissue fluid is electrolyzed to generate oxygen at the cathode and hydrogen at the anode. Therefore, the treatment will increase hydrogen level in vivo. Whether the therapeutic effects are correlated with the generation of hydrogen is worth exploring. Magnesium is used in the soluble biological materials such as bone material and blood vessel stents. When the material is dissolved, magnesium reacts with water producing magnesium hydroxide and hydrogen. Therefore, the effects of these biological soluble materials on diseases could take hydrogen into account.

In conclusion, major breakthroughs could be in these following areas:

1. The molecular mechanism study: identify the target molecule responsible for the effects of hydrogen.
2. Clinical studies: select the appropriate disease and conduct strict double-blinded randomized clinical trials. The key is to confirm the therapeutic effects. The promising candidate diseases are autoimmune diseases like rheumatoid arthritis, neurodegenerative diseases like senile dementia, and metabolic diseases like diabetes.
3. The hydrogen clinical use study: establish a reasonable clinical application scheme, acquire the access to clinical use as a drug, clarify the dose–response relationship, and evaluate the effectiveness of various routes of hydrogen administration.
4. The study of roles of hydrogen in a variety of biological systems such as in bacteria, fungi, plants, and so on. Although hydrogen certainly plays an important role in bacteria and protozoa, its effects on fungi and plants still lack evidence.
5. The study of other similar gas molecules: clarify the biological effects of some other simple intoxic gas molecules like methane and helium.

Many classic "small" molecules, like small peptides, can take effects only by contacting other molecules and, like vitamins and metal ions, serve as coenzymes only by binding to certain points of macromolecules. In the microenvironment of the biological fluid, the gas distribution is either in an aqueous phase or in a lipid phase. More importantly, it can enter into the interior of biological macromolecules, which is not possible for many macromolecules. Perhaps the secret of biological effects of these gases is just hidden within the simplest chemical properties.

References

1. Dole M, Wilson FR, Fife WP. Hyperbaric hydrogen therapy: a possible treatment for cancer. Science. 1975;190(4210):152–4.
2. Gharib B, Hanna S, Abdallahi OM, Lepidi H, Gardette B, De Reggi M. Anti-inflammatory properties of molecular hydrogen: investigation on parasite-induced liver inflammation. C R Acad Sci III. 2001;324(8):719–24.
3. Ohsawa I, Ishikawa M, Takahashi K, Watanabe M, Nishimaki K, Yamagata K, et al. Hydrogen acts as a therapeutic antioxidant by selectively reducing cytotoxic oxygen radicals. Nat Med. 2007;13(6):688–94.
4. Fukuda K-I, Asoh S, Ishikawa M, Yamamoto Y, Ohsawa I, Ohta S. Inhalation of hydrogen gas suppresses hepatic injury caused by ischemia/reperfusion through reducing oxidative stress. Biochem Biophys Res Commun. 2007;361(3):670–74.

5. Hayashida K, Sano M, Ohsawa I, Shinmura K, Tamaki K, Kimura K, et al. Inhalation of hydrogen gas reduces infarct size in the rat model of myocardial ischemia-reperfusion injury. Biochem Biophys Res Commun. 2008;373(1):30–35.
6. Buchholz BM, Kaczorowski DJ, Sugimoto R, Yang R, Wang Y, Billiar TR, et al. Hydrogen inhalation ameliorates oxidative stress in transplantation induced intestinal graft injury. Am J Transplant. 2008;8(10):2015–24.
7. Li J, Wang C, Zhang JH, Cai J-M, Cao Y-P, Sun X-J. Hydrogen-rich saline improves memory function in a rat model of amyloid-beta-induced Alzheimer's disease by reduction of oxidative stress. Brain Res. 2010;1328:152–61.
8. Nagata K, Nakashima-Kamimura N, Mikami T, Ohsawa I, Ohta S. Consumption of molecular hydrogen prevents the stress-induced impairments in hippocampus-dependent learning tasks during chronic physical restraint in mice. Neuropsychopharmacology. 2009;34(2):501–08.
9. Ohsawa I, Nishimaki K, Yamagata K, Ishikawa M, Ohta S. Consumption of hydrogen water prevents atherosclerosis in apolipoprotein E knockout mice. Biochem Biophys Res Commun. 2008;377(4):1195–98.
10. Noda K, Tanaka Y, Shigemura N, Kawamura T, Wang Y, Masutani K, et al. Hydrogen-supplemented drinking water protects cardiac allografts from inflammation-associated deterioration. Transplant Int. 2012;25(12):1213–22.
11. Cardinal JS, Zhan J, Wang Y, Sugimoto R, Tsung A, McCurry KR, et al. Oral hydrogen water prevents chronic allograft nephropathy in rats. Kidney Int. 2010;77(2):101–09.
12. Kajiyama S, Hasegawa G, Asano M, Hosoda H, Fukui M, Nakamura N, et al. Supplementation of hydrogen-rich water improves lipid and glucose metabolism in patients with type 2 diabetes or impaired glucose tolerance. Nutr Res. 2008;28(3):137–43.
13. Xie K, Hou L, Wang G, Xiong L. Effects of hydrogen gas inhalation on serum high mobility group box 1 levels in severe septic mice. Zhejiang Da Xue Bao Yi Xue Ban. 2010;39(5):454–7.
14. Xie K, Yu Y, Pei Y, Hou L, Chen S, Xiong L, et al. Protective effects of hydrogen gas on murine polymicrobial sepsis via reducing oxidative stress and hmgb1 release. Shock. 2010;34(1):90–97.
15. Xie K, Yu Y, Zhang Z, Liu W, Pei Y, Xiong L, et al. Hydrogen gas improves survival rate and organ damage in zymosan-induced generalized inflammation model. Shock. 2010;34(5):495–501.
16. Ji X, Tian Y, Xie K, Liu W, Qu Y, Fei Z. Protective effects of hydrogen-rich saline in a rat model of traumatic brain injury via reducing oxidative stress. J Surg Res. 2012;178(1):E9–E16.
17. Cai J, Kang Z, Liu WW, Luo X, Qiang S, Zhang JH, et al. Hydrogen therapy reduces apoptosis in neonatal hypoxia-ischemia rat model. Neurosci Lett. 2008;441(2):167–72.
18. Chen Q, Chen P, Zhou S, Yan X, Zhang J, Sun X, et al. Hydrogen-rich saline attenuated neuropathic pain by reducing oxidative stress. Can J Neuro Sci. 2013;40(6):857–63.
19. Ishibashi T, Sato B, Rikitake M, Seo T, Kurokawa R, Hara Y, et al. Consumption of water containing a high concentration of molecular hydrogen reduces oxidative stress and disease activity in patients with rheumatoid arthritis: an open-label pilot study. Med Gas Res. 2012;2(1):27.
20. Chen H, Sun YP, Li Y, Liu WW, Xiang HG, Fan LY, et al. Hydrogen-rich saline ameliorates the severity of L-arginine-induced acute pancreatitis in rats. Biochem Biophys Res Commun. 2010;393(2):308–13.
21. Li J, Wang C, Zhang JH, Cai J-M, Cao Y-P, Sun X-J. Hydrogen-rich saline improves memory function in a rat model of amyloid-beta-induced Alzheimer's disease by reduction of oxidative stress. Brain Res. 2010;1328:152–61.
22. Wenjie HAN, Min OU, Yuhong LIU, Qiang SUN. Protective effect of saturated hydrogen saline on hyperoxic lung injury. Chin J Naut Med Hyperb Med. 2011;18(3):129–32.
23. Sun Q, Cai J, Zhou J, Tao H, Zhang JH, Zhang W, et al. Hydrogen-rich saline reduces delayed neurologic sequelae in experimental carbon monoxide toxicity. Crit Care Med. 2011;39(4):765–69.
24. Sun H, Chen L, Zhou W, Hu L, Li L, Tu Q, et al. The protective role of hydrogen-rich saline in experimental liver injury in mice. J Hepatol. 2011;54(3):471–80.

25. Chen C, Chen Q, Mao Y, Xu S, Xia C, Shi X, et al. Hydrogen-rich saline protects against spinal cord injury in rats. Neurochem Res. 2010;35(7):1111–18.
26. Hayashi T, Yoshioka T, Hasegawa K, Miyamura M, Mori T, Ukimura A, et al. Inhalation of hydrogen gas attenuates left ventricular remodeling induced by intermittent hypoxia in mice. Am J Physiol Heart Circ Physiol. 2011;301(3):H1062–H69.
27. Zhang J, Wu Q, Song S, Wan Y, Zhang R, Tai M, et al. Effect of hydrogen-rich water on acute peritonitis of rat models. Int Immunopharmacol. 2014;21(1):94–101.
28. Kajiya M, Silva MJB, Sato K, Ouhara K, Kawai T. Hydrogen mediates suppression of colon inflammation induced by dextran sodium sulfate. Biochem Biophys Res Commun. 2009;386(1):11–15.
29. Chen X, Zhai X, Shi J, Liu WW, Tao H, Sun X, et al. Lactulose mediates suppression of dextran sodium sulfate-induced colon inflammation by increasing hydrogen production. Dig Dis Sci. 2013;58(6):1560–8.
30. Cai J, Kang Z, Liu K, Liu W, Li R, Zhang JH, et al. Neuroprotective effects of hydrogen saline in neonatal hypoxia-ischemia rat model. Brain Res. 2009;1256:129–37.
31. Sun Q, Kang Z, Cai J, Liu W, Liu Y, Zhang JH, et al. Hydrogen-rich saline protects myocardium against ischemia/reperfusion injury in rats. Exp Biol Med. 2009;234(10):1212–19.
32. Zhang Y, Sun Q, He B, Xiao J, Wang Z, Sun X. Anti-inflammatory effect of hydrogen-rich saline in a rat model of regional myocardial ischemia and reperfusion. Int J Cardiol. 2011;148(1):91–95.
33. Li H, Zhou R, Liu J, Li Q, Zhang J, Mu J, et al. Hydrogen-rich saline attenuates lung ischemia-reperfusion injury in rabbits. J Surg Res. 2012;174(1):E11–E16.
34. Shao-hua S, Xiao-yun S, Zhi-ren FU, Fang LIU, Wen-yuan GUO, Hong FU, et al. The effects of hydrogen-rich saline on renal ischemia/reperfusion injury in mice. Chin J Organ Transplant. 2010;31(2):109–13.
35. Zheng X, Mao Y, Cai J, Li Y, Liu W, Sun P, et al. Hydrogen-rich saline protects against intestinal ischemia/reperfusion injury in rats. Free Radic Res. 2009;43(5):478–84.
36. Penders J, Kissner R, Koppenol WH. ONOOH does not react with H: potential beneficial effects of H as an antioxidant by selective reaction with hydroxyl radicals and peroxynitrite. Free Rad Biol Med. 2014;75C:191–94.
37. Fujita K, Seike T, Yutsudo N, Ohno M, Yamada H, Yamaguchi H, et al. Hydrogen in drinking water reduces dopaminergic neuronal loss in the 1-methyl-4-phenyl-1,2,3,6-tetrahydropyridine Mouse Model of Parkinson's Disease. Plos One. 2009;4(9):e7247.

MIX
Papier aus verantwortungsvollen Quellen
Paper from responsible sources
FSC® C105338

If you have any concerns about our products,
you can contact us on
ProductSafety@springernature.com

In case Publisher is established outside the EU,
the EU authorized representative is:
Springer Nature Customer Service Center GmbH
Europaplatz 3, 69115 Heidelberg, Germany

Printed by Libri Plureos GmbH
in Hamburg, Germany